Common Sense Planned Maintenance

Clifford Jones, BCom, MSc.

For Linh.

ACKNOWLEDGEMENTS

I greatly appreciate the colleagues and senior leaders in HEINEKEN who have supported me in publishing this book and have allowed it's publication, believing that it will help the beverage industry and Engineering students to follow a Common Sense approach to Planned Maintenance.

Additionally I would like to thank:

Akshay Setlur Ramamohan for your enthusiasm for Planned Maintenance and your energetic support of the HACS system and its development (and for proofreading the early drafts) you are an outstanding Engineer in every way.

Jan Paul Boon for realizing the need to have Planned Maintenance systems, for sponsoring the HACS project, for trusting me to build the system and for remaining its champion across the globe.

Kubatbek Alimbekov and **Nguyen Truong Giang** for forming the original Steering Committee with me, and for developing the concepts and approaches that made the HACS the success that it is. You helped us to understand the need for the hierarchy, the need for construction groups and many practical concepts that are missing in other RCM approaches.

Dennis van der Plas, as leader of the Heineken Global Maintenance team for promoting the HACS Planned maintenance approach, always acknowledging the work we have done and building on it constructively. Also, for proofreading the early draft and your invaluable support in getting permission for publication.

Dr Jun Jie Chong at Newcastle University in Singapore, for encouraging his students to work with us on the HACS as interns, and for including my RCM lectures in his Syllabus every year (and also for proof reading).

Adam Spencer for sowing the seed that started this book, for your unflagging support of the HACS and your outstanding work in building the HACS on-line training modules.

Vivien Chan Hui Ling & Joycelyn Foo Pei San for your groundbreaking Capstone project work updating the findings of Nowlan and Heap.

Finally, a warm thankyou to all of the **SIT students and HEINEKEN Engineers** who worked on the HACS files and put up with my demands to follow the patterns and conventions that we developed, I hope that you all enjoyed the process and that together we became better Reliability Engineers.

DISCLAIMER

This book represents the personal views of the author alone.

This book describes an approach to developing Planned Maintenance schedules that has been applied in the beer manufacturing industry. This approach focuses on building a Planned Maintenance system where the assets have already been in operation for some time, and the same assets exist at many locations. A different approach may be needed for unique innovations or unfamiliar assets.

This approach may not be suitable for industries with higher inherent risk profiles or consequences of failure, such as the aerospace, power generation, petro-chemical or pharmaceutical industries.

This book is published only to assist Engineers, students and maintenance practitioners in the beverage industry.

Although the author has applied his best endeavors to ensure that this book is as accurate and correct as possible, he does not guarantee that this is completely accurate and correct, and thus cannot be relied on as such. Neither the author nor HEINEKEN NV or its subsidiaries assume any responsibility whatsoever in respect of or arising out of or in connection with the content of this material and third parties. If any third party chooses to use the contents of this book then he/she accepts and approves to do so entirely at his/her own risk.

ARTIFICIAL INTELLIGENCE

No artificial intelligence software was used to write or improve the content of this book. It was all written by a real person.

TABLE OF CONTENTS

FOREWORDS

Dr Jun Jie Chong, Assistant Professor, Mechanical Design and Manufacturing Engineering, Newcastle University Singapore:

Over the few years that I have known Clifford, he has consistently demonstrated a deep passion for sharing his knowledge and expertise with others, particularly in the field of maintenance engineering. He understands the importance of cultivating the next generation of engineers with the necessary maintenance skillset and has made it a priority to mentor our students and train in our undergraduate program (SIT-Newcastle, Mechanical Design and Manufacturing Engineering).

His dedication to education and knowledge-sharing is truly inspiring and has helped to shape the careers of countless individuals in the maintenance industry. In addition, it was clear that writing a book on the subject was a natural next step, as it would provide a platform for him to share his insights with a wider audience and help to advance the industry as a whole.

Receiving the draft of his book was an incredible privilege, and I felt truly fortunate to have been entrusted with such a valuable resource. As I delved into the pages, I was struck by his depth of knowledge and expertise, and the clarity and accessibility of his writing. It was clear that he had poured his heart and soul into the book, and that he was dedicated to providing the most comprehensive and informative resource possible. When he invited me to write a foreword for him, I felt humbled and honored to have the opportunity to contribute to his work in some small way. It was an opportunity to show my appreciation for his dedication and commitment to the field of maintenance, and to share my thoughts and insights with others who are also passionate about this important subject. I hope that my review will do justice to his incredible work, and that it will inspire others to learn more about the critical role that maintenance systems play in driving business success.

The book provides a comprehensive overview of planned maintenance and how it can be applied to improve equipment reliability and optimize maintenance processes. Cliff draws on his extensive experience in the brewery industry to provide practical advice and real-world examples of maintenance in action. He explains how Reliability Centered Maintenance (RCM) can be used to identify and prioritize maintenance tasks based on their impact on equipment reliability and overall business performance.

One of the prominent features of his book is the detail, from initial planning and scoping to data collection, analysis, and implementation. He provides detailed guidance on each step of the process, outlining key considerations, potential challenges, and best practices for success to improve asset reliability, reduce downtime, and lower costs. The inclusion of numerous photos and examples helps to illustrate key concepts and makes the guidance more accessible and easier to read.

Cliff is able to explain technical concepts in a way that is easily accessible to non-technical readers. For example, he uses an electric kettle, an item that nearly everybody is familiar with, to explain how to conduct failure mode and effect analysis (FMEA).

Overall, his book is an invaluable resource for anyone who wants to learn more about maintenance and its practical applications. With its detailed guidance, practical insights, and wealth of examples, it provides a comprehensive and accessible introduction to this critical aspect of maintenance engineering.

Dennis van der Plas, Manager Maintenance and Asset Care, HEINEKEN:

I have worked with Clifford with frequent personal interactions for over 10 years.

Currently I lead the Global Maintenance team, a small central team based in Holland, operating within our network of Operating Companies (OpCos), Regional Hubs and Centres of Excellence (CoEs), topic-based Communities and implementation task forces. My team is responsible for improving the Maintenance capability of the Heineken company and we do that by identifying the best practices, standardise these and roll out in programs, together with our subject matter experts around the world.

Clifford and his team are front runners in adopting new ways of working and showcasing what should become a global standard. Last year, when I visited them in Cambodia, it was great to see how the total capability approach for Maintenance works in practice and delivers business value. I use the learnings of my visit frequently to explain to others what the potential is of a well working Planned Maintenance capability.

I was fortunate to participate or have a leading role in many of the global Maintenance programs that are described in this book, such as Scale, the Sahara program, Supportability Analysis, TPM Next, Maximo and Asset Care, and to see the transformation of the Maintenance function in such a large company and all the benefits it brings.

Building a company-wide Maintenance capability is a journey, with many initiatives, packed into global programs (some more successful than others). It is a journey that's never finished as new technology keeps providing new opportunities, while we need to sustain the basics at the same time. I believe Clifford provides a very realistic view of the different initiatives that have been launched over the last decade in Heineken, showing the strengths and weaknesses of each initiative in such a way that the reader can learn from it.

Clifford describes very well how the Sahara program was the first capability program for Maintenance in Heineken, with a clear focus on organization structure and work routines. The program focused on restoring basic conditions before attention could turn to developing Planned Maintenance Schedules, and building this foundation is very necessary. The first global Reliability Centred Maintenance (RCM) attempt, the Supportability Analysis, had aimed to develop the Planned Maintenance schedules. Although the standard (Maintenance Standards Teams Route, described in this book) was well intentioned, the project

3

failed in execution. This failure was due to several causes, including: 1) high turnover in the people leading the initiative 2) too much focus on methodology and filling the templates, instead of creating scalable and practical work instructions, and 3) most important, the analysis was not done by brewery teams, resulting in impractical standards and a lack of ownership.

The development of the Heineken Asset Care Standards (HACS) avoided the mistakes of earlier attempts and resulted in an RCM program with a database of very suitable and practical standards that are now at the heart of the Asset Care program in Heineken, which is now being rolled out in 40 countries.

At the moment of writing this foreword, I received the news that Clifford with his team in Cambodia received the TPM Bronze Award, meaning that there's a solid foundation for Operations and Maintenance and the further development of the brewery will now focus on advanced methodologies. Knowing Clifford, he will explore the possibilities and showcase their implementation in practice. Having just finished this book, I'm already looking forward to the next one!! Will that be about Predictive Maintenance, making use of Artificial Intelligence? Remote Maintenance without site visits from specialists? The power of a global collaborative network? Total Asset Management? Autonomous robots doing inspections and even executing tasks? We can already see the developments happening around us, but the big question is "How does it work in our industry and how can we roll it out effectively?".

The foundation that Clifford described in this book will be the perfect starting point to continue the journey and the practical considerations will also apply to future developments. I will be very happy to work with Clifford again on this journey.

I found this book both entertaining and a good source of knowledge at the same time. Clifford alternates throughout the book easily between explanation, analysis, anecdotes, and instruction, which makes the book very readable. I believe the book is appealing for both young engineers, as well as senior Management in Maintenance, because it describes the basics of Maintenance Engineering as well as the considerations one needs to make when starting a large program. Altogether, I'm certain this book is an invaluable resource for many people.

CHAPTER ONE: INTRODUCTION

In this book I have described how I designed, developed and implemented the Planned Maintenance system that is being adopted throughout the HEINEKEN company, moving production operations from frequent breakdowns with expensive corrective maintenance and ineffective overhauls, to simultaneously reducing both maintenance costs and the number of breakdowns by applying a Common Sense Planned Maintenance system.

The system is called the HACS, the HEINEKEN Asset Care System, it is based on Reliability Centered Maintenance (RCM) theory, but with a common-sense approach that is appropriate for our industry and is derived from my 40 years of experience in operating and maintaining brewery equipment.

Whilst the HACS contains about 10 000 different Planned Maintenance schedules overall (a single brewery asset (machine) needs about 100-200 different Planned Maintenance schedules to maintain its condition indefinitely), it was built with minimal resources over only 4 years, and without using any external consultants. Instead, we utilized a number of Trainees, Interns from Singapore Institute of Technology and our own Engineers to assist in writing the Planned Maintenance schedules.

In the breweries where the HACS has been fully rolled out, we have achieved a significant reduction in maintenance costs combined with a much lower rate of unplanned downtime. For example, the annual maintenance cost of our brewery in Cambodia is shown in Figure 1 after the HACS was implemented in 2019/2020.

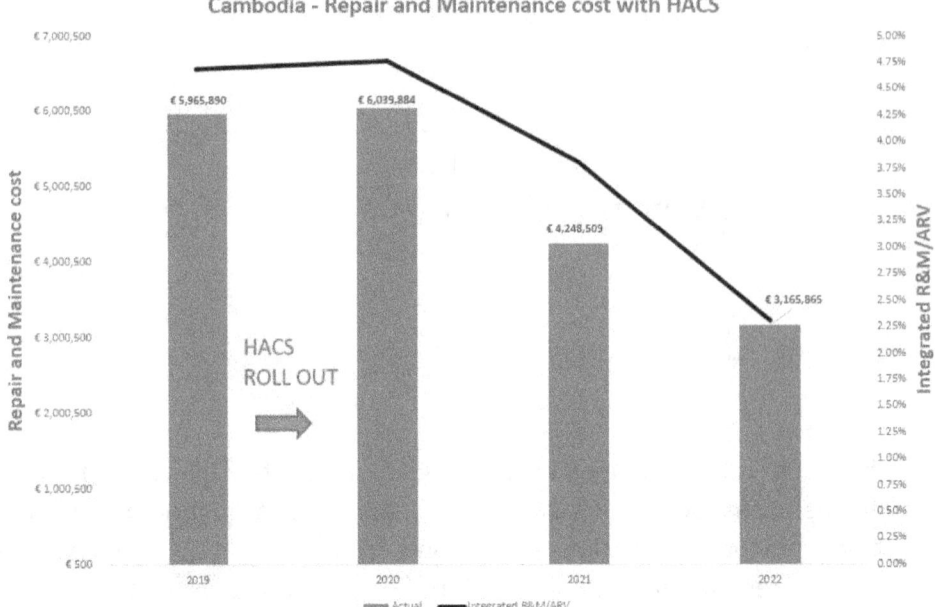

Figure 1: Cambodia brewery maintenance costs showing the effect of HACS implementation from 2019/2020.

The HACS system is designed to avoid tens of millions of dollars of investment in replacement capex and to deliver breakdown rates of 5% or lower in our operations where it is fully rolled out.

In this book you will find all of the theoretical AND practical information that you need to implement your own Common Sense Planned Maintenance system, to carry out Failure Mode and Effect Analysis in the real world and to write good clear Planned Maintenance schedules, Job Plans, and SOPs. I have included many actual examples of these, to make this the only guide you need.

I have carefully studied the work of other authors on the RCM (Reliability Centered Maintenance) subject: Nowlan and Heap, John Moubray, Neil Bloom, Anthony Smith and Nancy Regan to name a few. Each have their own focus and approach to RCM, and they have some significant differences to each other in specific areas. I have referenced and analyzed their work throughout this book to try to provide a balanced and academically sound summary. My book builds on theirs, focuses on some significant practical areas that are not covered by any of them, and updates the implementation of RCM in Planned Maintenance into the 2020's, with a focus on shop floor execution that is not found elsewhere.

The dominant maintenance strategy in the Brewing and Beverage industry today is one of huge annual overhauls, usually preceded by a paid supplier visit. The supplier makes a long list of parts that may or may not be needed, an order is placed by the brewery, and when the parts arrive the packaging line or brewhouse is shut down for some weeks whilst a crew of Technicians flies in, tears it down and overhauls everything. This results in a very high cost and a very low performance directly after the overhaul.

This (obsolete) overhaul/teardown methodology was the way of working in the aerospace industry in the 1960's, before the development of RCM. It was abandoned in the 1970's and 80's with the publication of Nowlan & Heap's paper: Reliability Centered Maintenance (Nowlan, 1978), following which the US military and the airline industry started to apply RCM. It is astounding that the **Brewing and Beverage industry has not adopted RCM in any significant way and is decades behind in the application of Planned Maintenance and RCM**.

One reason for this is because of the different consequence of failures in the different industries. In the aircraft industry a failure can mean a tragic crash losing hundreds of lives, or in the petrochemical industry failures can cause an environmental disaster. In these industries RCM and Planned Maintenance are quite thoroughly applied.

But in the Brewing and Beverage industry the consequence of almost all failures is very low, usually only a loss in efficiency. Our processes are relatively safe, and where we have specific equipment that can have serious failure consequences, such as boilers, ammonia cooling plants or CO_2 storage tanks, these are protected by regular statutory inspections rather than an RCM based Planned Maintenance system.

The Brewing and Beverage Industry has not had the motivation to adopt RCM and Planned Maintenance in a strategic and organized fashion and has instead proceeded with some under-resourced and sporadic attempts that were not successful, some of which are described in this book. I suspect that other FMCG industries may be in the same state.

The failure of the Brewing and Beverage industry to adopt RCM and Planned Maintenance has cost untold millions of dollars over the past years that would have been saved with an RCM based Planned Maintenance system.

For example, when I became Supply Chain Director in Cambodia in 2021, we discovered 2 million USD of unused spares left over from previous overhauls. There was no control over these parts or any plan to use them, and obviously the suppliers have no motivation to change this way of working as it is very beneficial to them.

It is clear that we need to implement RCM based Planned Maintenance to reduce maintenance costs, and there are already several theoretical texts available, so why have we not applied those in the Brewing and Beverage industry?

Well, most of the available texts are from author's who have worked in the aircraft or nuclear power industries (Nowlan and Heap, Regan, Bloom, Smith).

Their approaches reflect the requirements of those industries, to pedantically consider every possible eventuality in every system, which is usually appropriate in those fields. In addition, they share very little information on how to actually write the maintenance schedules that are required to prevent the breakdowns. They all focus on the RCM methodology (Failure Mode and Effect Analysis and RCM worksheets) with scant advice on what to do on the shop floor.

This focus on FMEA and decision diagrams, with a lack of an organized hierarchical structure and lack of consideration on how to write the actual Planned Maintenance schedules and Job Plans is why I have written this book, to enable Engineers to build a Common Sense Planned Maintenance system that works on the shop floor.

Further, all the available RCM texts assume that every machine is unique and brand new to the Engineers. This may be the case for a new aircraft or complex project, but almost all of our equipment has been around for many years with only small incremental improvements made by the manufacturer's each year. The components of these machines are a limited number of different OEM (Original Equipment Manufacturer) devices: pneumatic cylinders, solenoid valves, pumps, motors, sensors, photocells etc. that are all found across many of our machines, so we do not need to keep repeating the same failure analysis over and over for the same OEM parts, we can just copy them.

What is needed is an efficient method of copying or transposing the Planned Maintenance schedules for those common parts, which is why in developing the HACS we also built a database of common components with their Planned Maintenance schedules and Job Plans.

If you have operated your equipment for some years, and you have equipment that is not unique, then most failure causes have already happened somewhere, and you can focus on the maintenance required to prevent those failures, instead of an extensive FMEA exercise to identify potential failures that may or may not happen in the future.

The key to the development of the HEINEKEN Asset Care Standards across so many breweries in such a short time, with limited resources, was in applying a common structure to each machine (the hierarchy) and developing the hierarchy in such a way that similar machines or identical components can be quickly completed by utilizing the Planned Maintenance Schedules from the common/similar sections or assemblies.

I have found that the Hierarchy is at least as important as any other step in the Planned Maintenance schedule development, and yet it receives no significant attention in any of the available texts on Planned Maintenance and RCM.

If you are in the beverage industry, or in other FMCG manufacturing industries, then this book may be what you need to help you with the practical construction and application of a Common Sense Planned Maintenance program. However, if you are in the petrochemical, nuclear, power generation pharmaceutical, mining, or aerospace fields, this is probably not for you.

Going forwards, I will assume that you are an Engineer with reasonable technical knowledge, but nevertheless I will try to make it an entertaining read.

Clifford Jones
2024

Clifford Jones

CHAPTER TWO: ABOUT THE AUTHOR

This Chapter is included for those who might be interested to know how my upbringing and early career gave me the insights to become an expert on reliability and how I developed my hands-on common-sense approach. All of the events described in this section had some influence on the final product, a Common Sense Planned Maintenance System.

I was born in Singapore in 1964 to British Parents. My father was a Chief Petty Officer in Her Majesty's Royal Navy (RN), at that time based in Changi, Singapore. He was responsible for the maintenance of the Navy's Coastal Defense flotilla of 5 Minesweepers which were supporting the fledgling state of Singapore against Communist rebels trying to smuggle guns and generally be disruptive in the peninsular.

The Minesweepers were powered by Napier Deltic Diesel engines, an 18-cylinder opposed piston engine with 3 crankshafts (Figure 2). The RN maintenance manual for this had beautiful, exploded views and cut away diagrams, and was a treasured possession when I was young. In the manual were my father's study notes including his answers to examination questions, such as explaining how to set the timing on the engine's 3 crankshafts, one of which turns backwards.

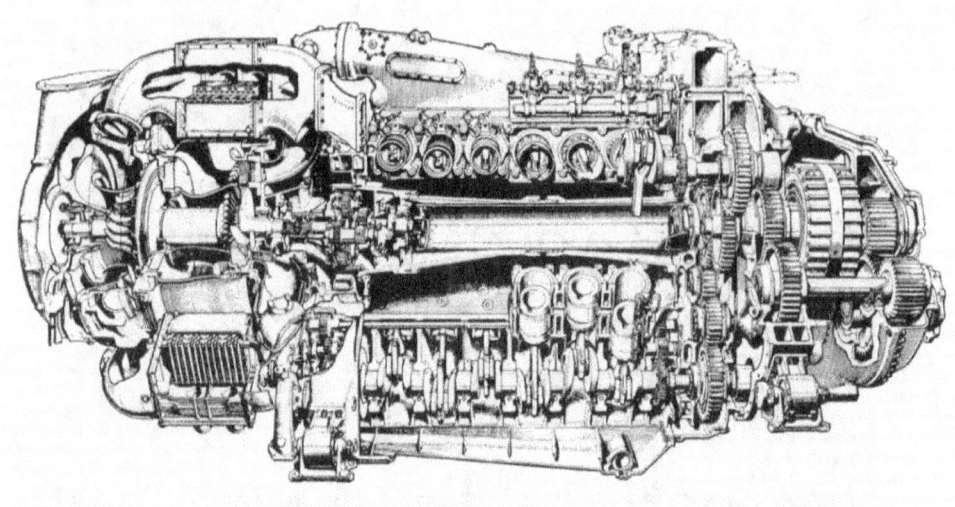

Figure 2: Cut-away of the Deltic Diesel engine, from the excellent article by William Pearce on the website Old Machine Press : https://oldmachinepress.com/2019/09/05/napier-deltic-opposed-piston-diesel-engine/ reprinted with permission.

Each minesweeper had 3 of these engines, 2 for propulsion and one connected to a pulse generator, a device that sent a huge electrical pulse down a cable stretched between 2 minesweepers, with the objective of exploding any magnetic mines near to the cable. According to my father, the pulse generator caused so much violent stress on the engine that they would have to be replaced every 3 or 4 months, whereas the identical propulsion engines had a 2000-hour lifespan before overhaul.

After serving in Singapore my father was posted to Gosport and served on HMS Brave Swordsman (Figure 3), which was at the time the fastest naval vessel in the world, powered by 3 Bristol Proteus gas turbine engines to a maximum speed of 96km/hr. To clean the blades of the gas turbines they would eat walnuts in the mess, at the end of the evening collecting and crushing the shells. The crushed shells were later thrown into the running gas turbine, there would be a lot of noise and black smoke, after which the turbine blades would be highly polished and look like new!!!

The Brave Swordsman was so fast that at full throttle the crew had to be strapped into their seats due to the violence of the impact with the waves.

Figure 3: HMS Brave Swordsman, powered by 3 Bristol Proteus gas turbines, note the additional fuel tanks amidships, next to the standing crewmen. From Lutz Walter collection.

After my father left the Navy, he started an Engineering company producing food processing machinery in King's Lynn in Norfolk, where I was brought up. I would visit his factory on a weekend and explore the workshop where the machines were built. I learned about fabrication from watching his employee's using cutters, grinders, welding equipment, lathes and shaping machines, and was fascinated by the process of manufacturing a machine from pieces of raw steel.

At home our garage had a workbench and tools, including some special tools that my father had made by hand as an apprentice (that I still have). Here I was encouraged to use the tools (as long as I put them away afterwards) and I was often covered in oil and grease on a cold winter's night, repairing my bicycle so that I could complete my paper round early the next morning, after which I had a 3-mile ride to school, regardless of the weather.

A treasured possession that my father bought for me was an Encyclopedia of "How things Work", which illustrated many mechanisms and devices that I spent hours studying, including items such as thermostats or one-way valves.

One day (I guess I was about 13) an employee of my Dad's, "Big Mick", delivered in his van a bare frame from a BSA motorbike, 2 wheels and a large box of parts that my Dad had bought from him for me (Figure 4).

Figure 4: BSA frame and Honda twin cylinder project. From author's own collection.

In order to assemble the parts, I first obtained a Haynes manual for the Honda 125cc twin engine. In those days the Haynes manual was an important tool, as the internet had not been invented yet, so it was the only source of information on how to overhaul/rebuild a vehicle as well as what settings were required. I assembled the engine using what skills I had: with limited cash, gaskets were rarely replaced, just refitted with more gasket compound. The correct torque for the cylinder head was estimated as I could not afford a torque wrench.

Somehow, I managed to start the completed engine. Big Mick welded up some engine mounts in my dad's workshop and with chain and sprockets installed we were ready to go. But there was very little power. I would push the bike up and down the road, bump-start the engine and it would run just a few feet before the engine would fade and die.

The issue was that there was no alternator/generator to charge the 6V battery, which itself could hold very little charge, the engine power was so low because the spark plug coil was being energized directly from the battery.

Further examination revealed a magnet on the crankshaft that rotated inside an alternator coil, which had a handful of wires that were not connected to anything. From the wiring diagram I figured out that this should produce AC current, and that I needed something called a rectifier to convert that to DC to charge the battery. So, I bought 4 diodes and soldered them to a board to make a diode bridge rectifier which was then connected to the battery. Problem solved, I hoped.

Alas I spent many more hours pushing and kickstarting the bike to try to start it, charging the weak battery to try again, but I never really managed to get it to run with any power.

What I learned from that experience was that I have a good intuitive understanding of mechanical equipment, but electronics is my weak point (but I did at least learn one way to convert AC to DC). Later I took some electronics classes at the local Technical College, but I have never managed to solder a joint that was not either dry or else melted through the circuit board.

Although my parents were fairly well off, they made me earn for myself the money that I needed for tools and maintenance, petrol, cars or social activities. At age 16 I was stacking shelves in the local Supermarket on the weekend, and also earning money with a summer job.

But thanks to my upbringing and the opportunities that my father created for me, I have the practical skills needed to use all types of mechanical tools, all of which has made it possible for me develop Planned Maintenance schedules and write Job Plans that a Technician can follow.

Throughout these formative years my father built his engineering business and also pursued a keen interest in flying; he used his light aircraft for business purposes as well as for pleasure. Almost every weekend I would visit the local flying club in Fenland, Lincolnshire, with him and we would perhaps do circuit practice or on some weekends fly to France or the Channel Islands. I was indeed very fortunate to have this incredible experience.

Sensible pilots (who want to reach old age) study the mishaps and accidents of others, so from an early age I read with my father just about every global aircraft accident report, either from the NTSB, CAA or FAA, usually published in the monthly aircraft Magazines "Flight" and "Pilot" that my father subscribed to. I have quite an extensive knowledge of the causes of many different aircraft accidents over the years, and the insights from this have also shaped my approach to Planned Maintenance.

One night on a flight from Heathrow back to our base in Norfolk, in our single engined Piper Comanche (Figure 5), we watched with growing dread as the engine oil pressure slowly dropped to zero and waited for the engine to fail leading to a dangerous forced landing in the dark. Happily, the engine continued to run normally, and we landed safely. Later the Maintenance Engineer found that the feed line to the oil pressure gauge was blocked by a small piece of rubber that had detached from a flexible seal.

Flexible seals account for a large proportion of the maintenance issues in process industries as we rely on them to regulate liquid or gas flow, but they are very prone to perishing and deterioration with time, temperature and wear. I learned a deep mistrust of flexible seals from this incident!

Figure 5: My Father at Fenland airfield with our 4-seater Piper Comanche 250, a very high-performance design at the time. From author's own collection.

Once I reached the age of 17, I purchased a Ford Escort Mk2 from my sister which had been in the family for some years. It had had almost no maintenance and was somewhat abused, but it was my first car. With a nearly indestructible 1300cc engine and drive train, most of my time was spent on bodywork repairs. Many a weekend I would be lying under the car in puddles of freezing water, filling holes in the door sills with fiberglass and painting over it with thick underbody paint. Somehow, we would get through the annual MOT test (arranged through big Mick), but in truth the car was more fiberglass and body filler than it was original steel.

At the same time, I was an active cadet in the Air Training Corps, attending training twice per week where we learned to march in formation, shoot .303 rifles and on weekends we had flying experience in gliders and aerobatic Chipmunks at nearby Cambridge airfield. A highlight for me was a summer camp at RAF Coningsby, then the home of the Battle of Britain Memorial flight, where I spent a week polishing the windows and cleaning the interiors of Spitfires, Hurricanes and their Lancaster bomber. It was a great honour to make the smallest contribution to maintaining these icons that defended us against Nazi aggression. Later I undertook evaluation for a Flying Scholarship at the Officer and Aircrew selection centre at Biggin Hill, and Her Majesty's Royal Airforce agreed to put me through a Private Pilot's Training course that I attended at

Cambridge airfield, completing all of the flying training for my Private Pilot's License at the age of 17.

I was fortunate to experience many hours of flying as a teenager, both privately and in military trainers. I have never really felt afraid when flying, as I have always felt I have some chance of control over the situation, always being seated in a cockpit with the controls in front of me. However, put me on a fairground ride where my safety depends purely on the design and maintenance of a device that I am strapped into with no controls or possibility of influence, and I become extremely nervous. My friends cannot understand why I can enjoy high-G aerobatics and yet I will not get on a fairground ride with them.

When I was 18 our family moved to South Africa, and I decided not to pursue a career in flying but to see what opportunities there were in other countries, leaving the UK at a time when it was crippled by industrial action and a depressed economy (early 1980's). In Johannesburg I owned a Fiat 132 for a few years (that my father bought for me), followed by a Ford Cortina V6. Like their predecessors, these cars never saw a professional workshop, everything that failed I had to fix myself. I have spent a weekend under the Fiat removing the gearbox and rebuilding it through the night, and several times rebuilding the V6 engine of the Ford, either due to leaking cylinder heads or stripped teeth on the drive gear to the valve cams.

Driving through central Johannesburg, feeling the front end of your car suddenly collapse and then seeing your front wheel bouncing across the road is a memorable lesson in the importance of sufficient thread length being available for securing nuts to do their job.

In those times I always had a set of tools in the boot of my car, as well as a few key spares and rags, such was the corrective maintenance regime. It is very possible that a tendency to drive at maximum speed and accelerate to maximum revs also had an impact on the amount of component failures.

I commenced a BSc at the University of the Witwatersrand, where unfortunately I lasted less than 2 weeks (to the great disappointment of my father). The lectures mostly seemed to involve very complex calculus equations, I just could not grasp why I needed to find the anti-derivatives of a function, and how it could possibly help me to earn an income. A BSc was far too theoretical for me at that stage. (In later years I received an MSc from Brunel University, but there was a lot less calculus).

To redeem myself I joined the South African Breweries (SAB) as a Trainee, simultaneously studying at the Witwatersrand Technikon (the South African equivalent of a Polytechnic) for a Diploma in Industrial Engineering, whilst working for SAB in 6 month blocks until the Diploma was completed.

Industrial Engineering was life-changing for me. I learned about manufacturing processes, motors, engines, steel production and different types of production lines. Ergonomics was a topic that I found particularly fascinating, studying the man-machine interface and the best and most efficient way to arrange production on the shop floor. I made friends at college who were like-minded, and we spent weekends building a beach buggy out of a Volkswagen Beetle and shoe-horning a Ford V6 into a Volkswagen campervan. Our fingernails were nearly always black, and our hands covered in small cuts from slipped tools.

(The loss of the Polytechnics in the UK, as they have been converted to Universities since the 1990's, has greatly reduced the number of tertiary qualifications that provide practical knowledge about how things work.)

At the brewery as a Trainee, I was learning about Brewing and Packaging. My training in each department usually started with the Supervisor giving me a broom and telling me to call him when the floor was clean. I remember introducing myself to the Head Brewer, a very large Boer (Afrikaans farmer) by the name of Dries Henning. I told him I had come to learn about Brewing. He asked me if I had a pencil, I said yes. He asked me if I had some paper, I said yes. Then he said, "Then f**k off and don't come back here until you have drawn a plan of every pipe in the brewhouse". It took me some weeks, but thanks to Dries, whenever I am in a brewhouse I automatically follow the pipes from one vessel to the next with my eyes and I nearly always know what is connected where.

As a young line manager in SAB, I also learned about leadership and conflict resolution, in what was one of the most dynamic and challenging environments a Manager could ever have experienced, as South Africa brought an end to the Apartheid regime and stumbled its way into democracy.

By the time that Nelson Mandela was released from jail in 1990 I was based in Cape Town (still with SAB), a Unit Manager running a large continuous shift bottling line with a huge team of 116 operators across 4 shifts. (Line 2 in Newlands, Cape Town, was rated at 45000 bph with 750ml bottles, some 33750 liters per hour). In parallel to the external political process, internally we were going through a major change process, supported by external consultants, and we were determined to forge a new South Africa together. Our main agenda was to put racism behind us and to work together in the rainbow nation.

To show their political support our workforce would join one protest march or another at least once or twice per week. All of the remaining Supervisors, Managers and Technicians would then go to the shop floor and run the bottling line in their place. This carried on for several months and in this time I was the operator of the bottle Filler, the Labeler or sometimes driving a Forklift.

I experienced first-hand many of the difficulties that operators experience every day in their jobs: How hard it was to clean the labeler parts without damaging them, and how difficult it is to set up the parts accurately. How often conveyor chains jump off of a magnetic flex curve, how much damage is done by just a small bump with a 4-ton Fork-Lift, how hard it is to see which valve is underfilling on the filler when it is running at full speed.

As a result, with the help of the Engineering team, we built parts-cleaning baths and set-up jigs for labelers that I have since introduced into every brewery that I have Supervised (Figure 6), and implemented other tools to make their jobs easier.

Figure 6: labeler parts cleaning bath designed by the author to make cleaning easier and reduce parts damage: From author's own collection.

My operating experience gave me a different approach to engagement at the shop floor level. Too often we don't really appreciate the difficulties that our shop floor workers face every day, there is so much we can do to make their jobs easier and more efficient. This has also been the bedrock of my ability to coach and train in my career, as you cannot really tell someone how to do a job if you have not done it yourself.

The performance in the first weeks that the Management team was running the bottling line was very low, and the number of bottles underfilled or mis-labelled was shocking, but gradually we learned what our operators had to do to make the line perform.

Such hands-on experience is much more difficult to obtain today, as we are more conscious of the safety risks, as we should be. However, I always do try to make sure that the trainees assigned to me spend time on the factory floor and learn to operate and maintain the equipment themselves when it is safe for them to do so.

I managed the bottling line in Cape Town with the support of two exceptional Engineers: Ivan Hull and Dave Scholtz, who taught me the 5-Why process (Root Cause Failure Analysis, see Figure 7), and how to use it to eliminate a failure from re-occurring. They were not formally trained in RCM theory as far as I know, but they used the 5-Why process very effectively to develop Planned Maintenance schedules. They insisted that 10% of the operating hours were allocated to Planned Maintenance each week, and that Planned Maintenance did not start until the line was properly cleaned.

5 Why sheet													
Failure Mode	Potential causes										4M	Actions	
	Why (1)	Check	Why (2)	Check	Why (3)	Check	Why (4)	Check	Why (5)	Check		CORRECTIVE ACTION	Check	PREVENTIVE ACTION	Check
	MSI and Failure mode were not identified in HACS	Y									Method		Determine and implement optimal maintenance strategy based on RCM flow chart (ref. xx), update HACS
	There is a lack of PM/CILT execution	Y	Quality of execution is poor	Y	No or low quality instruction	Y					Method	Develop work instruction	
					Lack/unavailability of required PM tools	V					Material	To have PM tools in place to perform the task	
					Lack of required experience	V					Man	Close skill gap	
			PM/CILT not executed On Time In Full	Y							Method	Optimise PM/CILT execution process	
Failure happened before anticipated lifetime			The operational context is different (environment, running hrs, stress, materials handled,...)	Y	The operational context changed without proper analysis (Management of Change)	V					Method / Material	MOC procedure	

Figure 7: Part of a 5-Why root cause failure analysis worksheet, from TPM worksheet.

Trying to get the shop floor teams to complete the 5-Why after a breakdown is repaired is always a challenge. With production pressures the Operators and Engineers want to get the line running, and often the 5-Why is not received until a few days or even weeks after the breakdown. Out of frustration I introduced a rule that the line could not be started until the 5-Why is completed. Once I even removed the power lock-out key from a machine and took it upstairs to my office with the statement that I would return it when the 5-Why was received. This was more to make a point about the importance of the 5-Why than anything else, but still today I teach that the 5-Why should be completed as part of the breakdown, and that the line should not be started until the 5-Why is done. (5-Why failure cause analysis is described in detail in Chapter 10).

Despite the challenges of the political transition to the new South Africa, Technicians (who were then predominantly white) and black and coloured Operators worked well together to enhance the performance of the line.

So effective was the planned maintenance regime of Ivan and Dave that we were able to produce 1 million bottles in 24 hours on several occasions, an efficiency of 92.6 % which I think is unprecedented and possibly has never been repeated on a returnable bottling line of this size. This speaks greatly of the motivation and engagement of the teams in Production and in Engineering, and I was immensely proud to be the leader of this team. This was a continuous shift operation, 24hours per day, seven days a week, that I managed for about 4 years.

More than thirty years later and the 5-Why process is still a key component of the Planned Maintenance and continuous improvement process, and something that is used every day in our breweries.

I left SAB in 1996 and for the next 5 years I ran a small training company, with clients including SAB and other companies, training Operators to run beverage packaging lines in South Africa, Poland, Russia, Romania, Zambia, Kenya, Tanzania, Mozambique, Ghana and anywhere else that I was needed. My hands-on experience allowed me to write clearly illustrated operating manuals, and to train the operators in both the theory and the practical operation of the equipment.

Having completed a Bachelor of Commerce degree in South Africa (through distance learning whilst I worked for SAB) I then fulfilled my life-long dream of studying at Brunel University and completed a Master of Science in Packaging Technology. (The dream was based on my admiration of the achievements of Isambard Kingdom Brunel, who built dockyards, railways, steamships and numerous bridges and tunnels across the UK in the early 19th century).

I wrote a set of illustrated training modules, "Fundamentals of Packaging", that explained the principles of Pasteurization, bottle Washing, Labeling, Conveying, Filling and Seaming in simple terms for Operators and Technicians. These modules are still used today, and for 5 years I was able share knowledge and build competence in my client's teams, which is for me the most rewarding thing that I can do.

However, the market for my training packages was very volatile and the lean months were very stressful (not least from a financial perspective). I longed to get back to the security and stability of full-time employment, so I returned to the UK and started applying for suitable positions.

Eventually I landed a job at Royal Brewery in Manchester as Engineering Development Manager. Scottish Courage needed someone to improve the relationship between the Craft Technicians, the Operators and Management, and they felt that my experience in South Africa in the final days of Apartheid had made me tough enough to survive working in Moss Side. This was probably the easiest job I ever had, as all the Craft Technicians wanted was Management to listen to them, and the Operators responded very positively to receiving my Fundamentals training. Much of the reason I succeeded in this role was my hands-on experience.

Unfortunately, I had only been in the role for 18 months when Scottish Courage closed 2 out of its 6 breweries with consequent headcount reductions so I needed to find better prospects. I moved to Pepsi in Saudi Arabia as the Production Manager of a large factory in Khamis Mushait, close to the border with Yemen. We had our own distribution fleet, so I learned a lot about maintaining trucks that I had not been exposed to before, and about making carbonated soft drinks. Living in Saudi Arabia at that time came with the constant threat of kidnapping and decapitation for expatriates, so I was fortunate to live on the UK-run air force base where we were fairly well protected.

In 2004 I joined HEINEKEN, initially starting as Brewery Manager in El Obour in Egypt, after which I moved to Nigeria as Supply Chain Manager. From Nigeria I moved to Vietnam as Supply Chain Director, and then to a regional role as Area Supply Chain Director based in Singapore, which I held for 8 years.

In the regional role I was responsible for our breweries in Mongolia, Laos, Cambodia, Thailand, Myanmar, Papua New Guinea, The Solomon Islands, New Zealand, New Caledonia, The Philippines, Sri Lanka and Timor Leste. In all of these locations I improved safety and quality standards, reduced losses and coached and supported the local teams. I designed and led the project to build the brewery in Timor Leste, the first industrial facility of any kind in that country, and I completely upgraded our breweries in the Philippines. However, the travelling was relentless, and I was away most weeks only returning home for the weekends. During this time my lovely wife gave birth to our two sons, somehow she managed to cope brilliantly with caring for two infants whilst I was frequently travelling overseas.

I also built the HEINEKEN Asset Care System (HACS) whilst in the regional role (see Chapter 5.2). A big reason for my asking to run the brewery in Cambodia as my next posting in 2021 was to be able to roll out the HACS and to prove its effectiveness (Cambodia is ideal for this as it is a large and modern operation and was the location where much of the HACS development work was done). I also wanted a job where I could stop travelling and be at home for my growing boys and long-suffering wife.

To demonstrate the effectiveness of a Planned Maintenance program there is nothing as convincing as showing the results in a real operation, and the results in Cambodia have contributed greatly to the speed at which the HACS is now being adopted globally. With the close co-operation of the Global Maintenance

Team, I am able to support the Planned Maintenance agenda for our whole organisation.

I am often asked by students or trainees what drove my career, what made me want to be a senior Manager? I can honestly say it was a desire to be a better leader than those that I was working under. To use my understanding of the shop floor to make better decisions, to seek the input and knowledge of the workforce to make operations more efficient. I cannot count the times that I see inefficiency on the shop floor just because we do not provide the right tools for the operators to use, or we do not understand the challenges that they face.

As a Supply Chain Director of a large brewery, I have the most opportunity to make those better decisions, to lead a team that is engaged and empowered, to coach and to train, and to shape our Brewery into a real learning organization.

In recent years I have coached and trained many Engineering students as Interns working on the HACS development and I also lecture in Singapore for the Singapore Institute of Technology/Newcastle University in Singapore. I find that many of today's students lack the hands-on practical knowledge and experience that I benefited from as a youngster. A good hands-on foundation is necessary in order to understand failure modes and how to prevent them and to be able to apply this when developing Planned Maintenance schedules. Outside of my Heineken responsibilities, I want to find more ways to train young Engineers in Reliability Engineering.

Clifford Jones

CHAPTER THREE: THE DEVELOPMENT OF RCM

3.1 THE EARLY YEARS

From the industrial revolution onwards (approximately 1780 to 1840), more and more complex machines were conceived, designed and built: the steam engine, the spinning Jenny, the power loom and the cotton gin to name a few. These machines were usually rugged, and being purely mechanical were simple enough for maintenance by the builder, by Mechanics or by Blacksmith's who had a good mechanical knowledge (Smith, 1993, p. 43).

At the start of the 20th Century, and again after World War 2, there was a great expansion in the number of manufacturing facilities across the world. Mass production was pioneered by Henry Ford in the 1920's, and the second World War led to the development of complex and powerful engines, including jet engines, the early rockets and the first computers. The proliferation of machines meant that the original builder was no longer the one carrying out maintenance or repairs.

Though much more complex, this machinery was still fundamentally mechanical, and where there were maintenance programs, they were developed by Mechanics or Engineers without much analysis and *based on the assumption that every component has a certain lifetime after which it must be overhauled, inspected and if necessary, replaced* (Take note of this assumption, it plays a big part in the development of RCM). Basically, it was assumed that parts will wear out after a certain amount of use. The more the hours of use, the more the wear.

But as the early computer systems were introduced, based on Vacuum valves (See Colossus in Figure 8) it was also observed that failures occurred at a high but decreasing rate for new components (Infant Mortality, Chapter 6.4), followed by a certain service life after which there was again a sharp increase in failure rate (wear out phase).

Figure 8: The Colossus, in 1943 the world's first programmable digital computer, from www.Britannica.com

This thinking is described in the failure pattern known as the Bathtub Curve (Figure 9), which is the shape of the curve when we plot component lifetime horizontally and frequency of failure vertically.

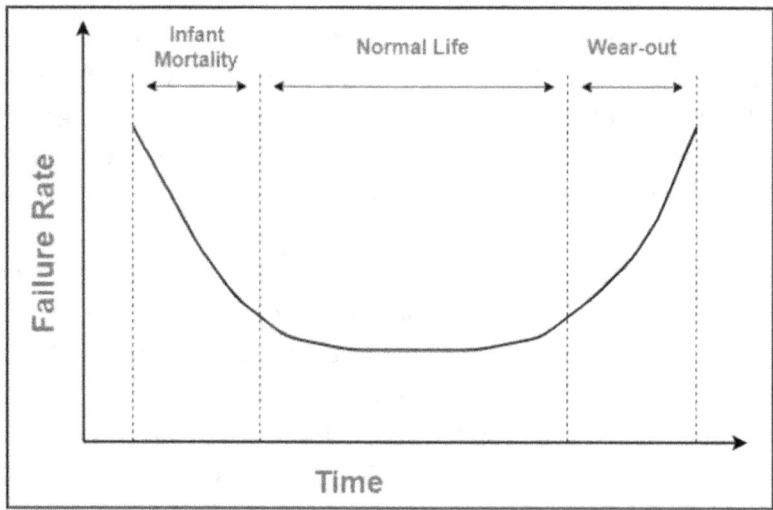

Figure 9: The Bathtub failure curve, the failure pattern that was assumed to apply to the majority of equipment components for many years.

In the 1950's and 1960's commercial jet aircraft were introduced, notably the Boeing 707 in 1957, Sud Aviation Caravelle and the Boeing 727 in 1958, followed by the Boing 737 in 1968.

Initially the aircraft manufacturers developed the maintenance programs for these aircraft themselves, but they were still based on the concept of overhauling every component after a set time, assuming that all components have a failure pattern that followed the bathtub curve.

At this time the rate of fatal Commercial aircraft accidents was very high, and as aircraft became more complex, the cost to the operating airlines of frequent extensive overhauls was a major concern.

The response to the increasing failures of this more complex equipment was to specify more frequent overhauls/teardowns.

Regan claims (Regan, 2012, p. 6) that there were 60 crashes per million takeoffs in the 1960's, two thirds of which were due to equipment failure. This rate would equate to more than 2 commercial aircraft crashes per day in modern times!!

Seeing that more overhauls and teardowns did not improve reliability, in 1960 the airline industry formed a taskforce to investigate preventive maintenance, the Air Transport Association Maintenance Steering Group (MSG).

This led to the development of on-condition maintenance (inspection). MSG-1 (as it became known) was described in the handbook *"Maintenance Evaluation and Program Development"* produced by the MSG and the Federal Aviation Authority (FAA) primarily for the Boeing 747 (first flown in 1969, Figure 10). MSG-1 applied a decision tree process for ranking Planned Maintenance tasks, still assuming that the bathtub curve applied to failure probability. But in the foreword to the program, it was stated: *"After careful study the committee is convinced that reliability and overhaul time control are not necessarily directly associated topics; therefore, these subjects are dealt with separately."* The committee focused on guidelines for propulsion system maintenance and found that in many cases there was no effective form of scheduled maintenance (meaning that failures still occurred despite frequent overhauls).

Figure 10: First flight of the magnificent Boeing 747, February 9th, 1969, image from AP.

In 1970 MSG-2 was described in the document *"Airline/Manufacturer Maintenance Program Planning"*. This approach was process oriented and analyzed failure modes from the part level upwards. It was used to develop maintenance programs for the Lockheed 1011 and the DC10. But the underlying principle remained that all components have a lifespan after which they must be overhauled/replaced.

From the 1960's onwards United Airlines was involved in trying to co-ordinate what was learned from these activities and to define a generally applicable approach to the design of maintenance programs. F. Stanley Nowlan (Director Maintenance Analysis) and Howard F. Heap (Manager Maintenance Program Planning) led by Thomas D. Matteson (Vice president of Maintenance Planning) along with Bill Mentzer (Senior Vice President Engineering and Maintenance) started to analyze the reasons behind various equipment failures, using a database of similar components across different aircraft. Failure density distributions were developed from the component operating history, and the failure rate was derived as a function of time.

Their findings were applied in 1972 in contracts for the Department of Defense to the US Navy's Lockheed P3 Orion aircraft and the US Airforce's McDonald Douglas F4J Phantom in 1974. In 1975 the DOD directed that the MSG concept be labelled **Reliability Centered Maintenance** and that is to be applied to all military systems (Smith, 1993, p. 48).

In 1978 Nowlan and Heap published their paper, "Reliability Centered Maintenance" under contract of the Department of Defense, which summarized their extensive analysis and defined RCM as we know it today (Nowlan, 1978).

Based on this, MSG-3 *Operator/Manufacturer Scheduled Maintenance Development* was published in 1980 and is the standard used today by the airline industry, though of course frequently updated since then.

3.2 NOWLAN AND HEAP'S FINDINGS: AN OVERVIEW

From examining the causes of failures of specific components in aircraft, Nowlan and Heap discovered 6 failure patterns as shown in Figure 11: The Bathtub curve, the Wear-Out curve, the Fatigue pattern, the Initial Break-In curve, the Random failure pattern and the Infant Mortality curve. These are described in detail in Chapter 6.4.

They found that only 4% of the component's follow the traditionally accepted Bathtub failure curve, and in total only 11% of components show an aging characteristic that can be addressed with traditional overhauls (Patterns A, B and C in Figure 11).

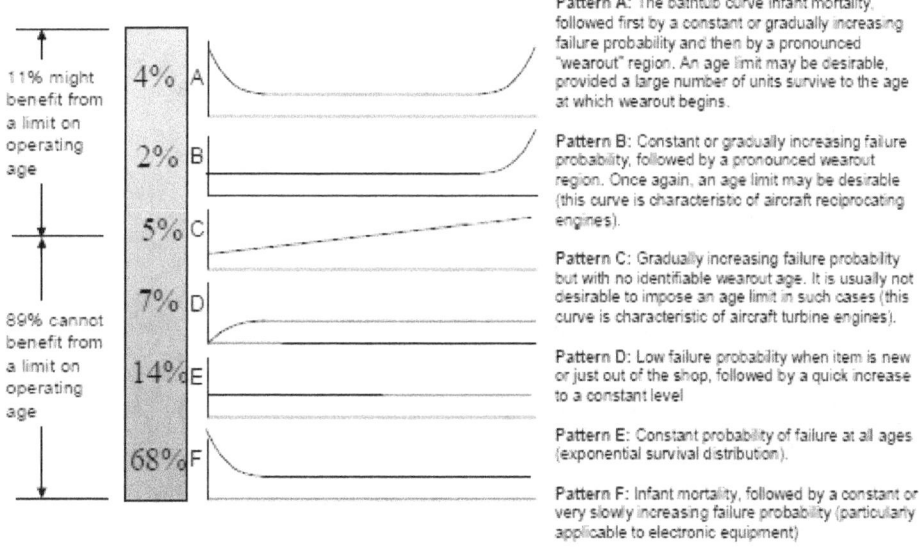

Figure 11: Age reliability patterns from Nowlan and Heap, Reliability-Centered Maintenance 1978.

They found 14% of components to have a constant probability of failure (Pattern E), meaning that any failure might be random (or at least the cause is not understood), and a whopping 68% of components that follow the Infant Mortality failure curve (Pattern F), meaning a high probability of failure immediately after installation.

Returning to the earlier statement that *"every component has a certain lifetime after which it must be overhauled, inspected and if necessary, replaced"* they showed this assumption to be incorrect. The strategy in the aircraft industry of frequent overhauls to prevent failures was shown to be appropriate for only 11% of failures.

Actually, the frequent teardowns were probably increasing the failure rate, as the overhaul introduces many new components, the majority of which follow an infant mortality failure pattern, so each overhaul would have introduced more failures than if they had never done the overhaul.

(Every Engineer in our industry knows that the lowest efficiency of a Packaging line or Brewhouse is the week after the annual overhaul. We look at this more in Chapter 6, but it is startling that the lessons learned in the aircraft industry in the 1970's **are not yet applied in the beverage industry**).

Back to Nowlan and Heap, the main points of their paper (Nowlan, 1978) are summarized here for you, as the original document is over 500 pages:

- The objective of the maintenance system is to realize the inherent reliability capabilities of the equipment for which they are designed, and to do so at minimum cost.
- Each scheduled maintenance task is generated for an explicit reason.
- The consequences of each failure possibility are evaluated, and the failures are then classified according to the severity of their consequences.
- A significant item is one whose failure could affect safety or have economic consequences.
- For all significant items (safety or cost) proposed tasks are evaluated according to specific criteria of applicability and effectiveness.
- The resultant maintenance program contains all the tasks necessary to protect safety and operating reliability, and only tasks that will accomplish this objective.
- Decision logic starts with an evaluation of the consequence of functional failure followed by an evaluation of the failure modes that cause the loss of function.
- There are 4 maintenance tasks defined:
  - On condition inspection.
  - Rework (overhaul) of an item before a specified age limit.

- - Discard (time-based replacement) before a specified age limit.
  - Failure finding inspections of a hidden function item.
- The role of a hidden failure in a sequence of multiple independent failures is stressed.
- A failure is an unsatisfactory condition.
- A functional failure is the inability of an item to meet a performance standard.
- A potential failure is an identifiable physical condition which indicates that a functional failure is imminent.
- The consequences of the failure determine the priority of the maintenance effort.
- The maintenance system requires an organized information system (nowadays we have CMMS, see Chapter 11).
- The program must be dynamic and respond to new data.
- The basic principles of RCM are:
  - How does a failure occur?
  - What are its consequences?
  - What good can preventive maintenance do?
- The process of developing an RCM maintenance program consists of
  - Partitioning the equipment into object categories to identify those items that require intensive study.
  - Identifying significant items whose failure would have a consequence for safety and cost, and all hidden functions.
  - Identifying the maintenance requirements of each significant item and hidden functions.
  - Deciding whether to modify the design or allow run to failure.
  - Selecting conservative initial intervals for each task.
  - Grouping tasks into a maintenance package.
  - Establish a program to provide factual information to revise the initial decisions.
- Failure is a process whereby a failure occurs when the amount of stress exceeds the remaining failure resistance. Failure resistance fluctuates over time.
- Reliability is the probability that an item will survive to a specified operating age, under specified operating conditions, without failure.
- RCM decision diagrams were provided to evaluate the consequence of failure and the appropriate maintenance task.

The remainder of the 500-page paper goes into many examples of the practical application of the above principles to specific aircraft maintenance programs, such as the McDonnell Douglas DC10.

You can download their paper at:

https://reliabilitywebfiles.s3.amazonaws.com/Reliability+Centered+Mainten ance+by+Nowlan+and+Heap.pdf

3.3 NOWLAN AND HEAP: STILL RELEVANT TO THE BEVERAGE INDUSTRY?

When we look at the findings of Nowlan and Heap, we have to view them from their starting point, which was that there was a massive excess of overhaul maintenance being carried out in the aircraft industry, with millions of man hours spent on unnecessary tear downs of aircraft or assemblies. They give the example that the structural inspections of the McDonnell Douglas DC-8 required an incredible 4 million man hours to complete for up to 20 000hrs of flight time, whereas for the Boing 747 the same inspections required 66 000 hours using the new approach to Planned Maintenance brought about by RCM (Nowlan, 1978, p. 6).

We will look at the status of the beverage industry and our breweries in later chapters, but our starting point is somewhat different. In general, we have the opposite situation, a lack of preventive maintenance, and an excess of corrective/breakdown maintenance.

There is a great focus in Nowlan and Heap on hidden failures, inspecting for them and their contribution to a sequence of multiple independent failures. Hidden failures refer to systems that are not usually used in operation (whose failure is not apparent to the operator), for example a fire detection system, a fire extinguishing system or a back-up generator.

This obviously has more significance in the aircraft industry where there is a lot of built in redundancy. Typically, there are 3 separate hydraulic systems for the aircraft flight controls, 3 separate navigation computers, 2 or more separate sets of key instruments like airspeed and attitude indicators, so that the failure of one item is not normally a critical event, but you don't want to discover that the backup system is not operating at the time that you need it most.

Bloom (Bloom, 2006, p. 34) explains in detail in his book how systems working in Parallel (2 pumps in parallel for example in a standby power supply system) can incur a hidden failure if one pump fails as the other continues to run and meet demand, leading to his use of the term Potentially Critical, in that one pump failure is not critical if the standby system is not in use, but it is critical if the standby system is in use and the other parallel pump fails. He defines a potentially critical component as one whose immediate failure is not evident and is not immediately critical but has the potential to become critical.

In the brewing and beverage industry we do not have the level of redundant or parallel systems that exist in the aircraft or nuclear power industries. It is possible for safety circuits on a machine's safety doors to have a hidden failure that is not detected (although they should always be designed fail-safe, so that if a component fails then the machine stops), but it was a simple matter to include functional checks in our periodic inspections that we have built into the HACS.

Check valves (one-way valves), safety valves, overload sensors and many other safety devices may not immediately affect the operation when they fail, so they could incur a hidden failure and be a potentially critical component. Rather than a separate treatment of hidden failures and potentially critical components, in the HACS we ensured that all systems that are not usually part of the operation are also included with the necessary Planned Maintenance schedules, i.e.: Safety valves, overload switches, emergency brakes etc....

The risk foreseen by Bloom and others is that safety devices are overlooked and allowed to run to failure because the failure is hidden. The structured hierarchical approach of the HACS eliminates the possibility that hidden failures are overlooked because we evaluate every MSI in the hierarchy.

In chapter 6 I review the findings of Nowlan and Heap in some detail, looking at the failure curves that they discovered and what these mean to our maintenance strategy to prevent failures. This analysis is not covered in depth in the RCM texts of other authors, and there is little linkage of failure patterns to maintenance strategy, which is something I have tried to improve on in Chapter 8.

In Chapter 7 I have described some very interesting research carried out by Singapore Institute of Technology (SIT) students under my supervision to assess how relevant Nowlan and Heap's findings are to the beverage industry today, taking into account the fundamental differences between the aircraft industry of the 1960's and beverage industry of the 2020's.

Clifford Jones

CHAPTER FOUR: RCMII, JOHN MOUBRAY AND OTHERS

4.1 RCM 2

From the 1980's John Moubray worked in industry implementing Reliability Centered Maintenance (RCM), which led him to publish his well-known text RCM2 (Moubray, 1991). It is still one of the standard reference texts referred to when designing a Planned Maintenance system. Shortly after Anthony Smith published his work, Reliability Centered Maintenance (Smith, 1993), Smith having been an acquaintance of Tom Matteson at United Airlines.

Moubray expanded on Nowlan and Heap's work (they had 3 questions) to specify 7 questions about the asset or system under review:

- What are the functions and associated performance standards of the asset in its present operating context?
- In what ways does it fail to fulfill its functions?
- What causes each functional failure?
- What happens when each failure occurs?
- In what way does each failure matter?
- What can be done to predict or prevent each failure?
- What should be done if a suitable pro-active task cannot be found?

DIFFERENT APPROACHES TO DEFINING FUNCTIONS AND FAILURES

Chapter 2 of Moubray's book is dedicated to describing functions, with a strong focus on the user, he seperates functions into primary and secondary functions.

He is very specific that the function statement must contain a verb, an object and a desired standard of performance. The standard of performance must be from the user's perspective, and Moubray explains in detail how to define performance standards in terms of the capability of the device and the operating context.

Smith (Smith, 1993, p. 28) defines reliability as the probability that a device will satisfactorily perform a specified function for a specified period of time under given operating conditions.

Smith focuses more on defining the entire system first (relating to a system within a nuclear facility, such as the cooling system), and defining the system boundaries and system interfaces. From this point he defines the functions and potential functional failures, and then builds a functional failure matrix listing the different functional failures that can occur on each equipment item, and from this matrix the failure modes are then listed and analysed.

Bloom goes to some length to explain why system boundaries and system interface definitions are not so important (Bloom, 2006, p. 21), and though he makes no direct reference, he seems, in my view, to be criticising the approach of Smith.

The SAE standard JA1011 defines what is an RCM process and includes the following seven basic questions (Bloom, 2006, p. 134):

- What are the functions of the asset ?
- What are the functional failures ?
- What are the failure modes ?
- What are the failure effects ?
- What are the failure consequences ?
- What are the Planned Maintenance tasks ?
- What must be done if a Planned Maintenance task cannot be specified ?

From the obvious similarity it seems that the SAE standard JA1011 is very much built on the work of Moubray.

DIFFERENT APPROACHES TO FMEA

Moubray, Smith and the others focus on building an RCM worksheet or Failure Mode Effect Analysis (FMEA) sheet (Chapter 9) from the perspective of defining functions and drilling down into each function to find the common causes of failure to deliver that function.

Bloom introduces his Consequence of Failure Analysis Worksheet (COFA) (Bloom, 2006, p. 95) which is similar to FMEA but he approaches FMEA at component level, although he does not go down to the level of the Maintenance Significant Item (MSI, Chapter 8.3). Conversely, Smith approaches FMEA at the system level.

In his Chapter 3 Moubray defines a functional failure as the *"inability of any asset to fulfill a function to a standard of performance which is acceptable to the user"*. He describes partial functional failures and defines how to record functional failures on an FMEA work sheet.

Chapter 4 of Moubray's book looks in detail at Failure Mode and Effect Analysis (FMEA), defining a failure mode as *"any event which causes a functional failure"*. Having listed the functions and the functional failures, the next step is to list every failure mode that might cause the functional failure. Moubray specifies that the description of the failure mode must contain a noun and a verb, (Such as "seized bearing").

Moubray goes into great detail as to how to analyze and describe failure modes, and then moves on to Failure Effects. He says: *"Failure effects describe what happens when a failure mode occurs"*. Here Moubray suggests we should record what evidence there is of the failure, how it poses a threat, how it affects production/operations, what physical damage is caused and what can be done to repair it.

He does mention "when listing failure modes, do not try to list every single failure possibility regardless of its likelihood".

It is the application of this statement that it very challenging. If you are in the aerospace industry, then probably you should try to list every single failure possibility. You don't know how those failures can contribute to other failures that might result in catastrophe. In 1989 an uncontained engine failure on a United Airlines DC10 caused the loss of all 3 hydraulic systems, despite a heroic landing with no flight controls 111 lives were lost, though 185 survived. This happened because the hydraulic systems were very close together in the tail section beneath the engine, and all 3 hydraulic circuits were damaged by the shrapnel from the engine's explosive failure. (Wikipedia).

https://en.wikipedia.org/wiki/United_Airlines_Flight_232

But in the beverage industry we have no such concerns. We can accept that some failures will still occur, and we might miss some failure modes. That is OK if we analyze the resulting breakdown and then eliminate this failure mode with an updated schedule in the Planned Maintenance system. (See Chapter 10 on 5-Why failure/root cause analysis).

Moubray mentions that the equipment manufacturers or other third parties are not usually able to provide all of the information required for the FMEA, and that the best source is the on-site Technicians or Operators, with which I fully agree.

SO WHAT TO DO IN THE BEVERAGE INDUSTRY?

Having reviewed the various authors, it was clear that there was not a single generally accepted approach as to how to partition and break down a machine for failure analysis, nor was it clear whether that failure analysis should be by system, top down or bottom up.

I looked at Moubray's Level of Analysis and Information Worksheet, (Moubray, 1991, p. 80) and found that it did not give the guidance we needed. Moubray states here that "One of the most common mistakes in the RCM process is carrying out the analysis at too low a level in the equipment hierarchy". Instead, he advocates to start the analysis at a higher level and to work down using systems and sub systems and identifying functions and failures at the different levels.

The fact that 3 different authors (Moubray, Smith, Bloom) have 3 different approaches as to how to break down the equipment and carry out the failure analysis highlights one of the challenges faced in implementing RCM, and is surely one reason why so many organisations fail

OUR SOLUTION: MAPPING THE HIERARCHY

What we decided to do is look at a whole system/machine/asset based on the Piping and Instrumentation Diagram (P&ID) for process equipment or using the Spare Parts Catalogue and maintenance manuals for Packaging machines.

We then follow a **hierarchical structure** to partition the machine in a structured way (Chapter 12) and continue the hierarchy to the level of the Maintenance Significant Item (MSI Chapter 8.3), **excluding all parts that are not MSIs**. Later we will examine each MSI and the physical change that can cause it's failure (the failure mode). Defining which parts are MSIs and which parts can be ignored from a maintenance perspective, is how we manage the volume of Planned Maintenance schedules required per asset to a reasonable level. (In our industry we can decide, for example, that a support bracket will probably never fail, and therefore for us it is not an MSI. In other high-risk industries, like aerospace, such a support bracket could be very important and would have to be included in the analysis).

This is the foundation of our approach in the HACS, and in Chapter 12, Building the HACS Hierarchy, I go into more detail as to how it is done.

This approach makes it easy to use the hierarchical structure of one machine to develop the Planned Maintenance schedules of another machine and is the enabler by which we are able to copy the Planned Maintenance schedules of the assemblies in one machine to another similar machine. This has allowed us, for example, to take a completed set of asset care standards for a Sidel can filler, and with minimum work adapt it to a KHS or Krones can filler. Without a structured hierarchy, development of the second machine would start from zero.

Our approach also takes into account the need to develop Planned Maintenance schedules for several sites with similar equipment and how to optimize that. Moubray and the others only ever seem to consider a machine or system in isolation.

As mentioned in the introduction, the components of these machines are a limited number of different OEM devices (OEM means Original Equipment Manufacturer, here it refers to parts the machine manufacturer obtains from other specialist suppliers): Pneumatic cylinders, solenoid valves, pumps, motors, sensors, photocells are all common OEM components that are found on many of our machines, from a very limited number of these suppliers, so if we have the Planned Maintenance schedules and Job Plans for a particular solenoid valve, we can just copy it everywhere that the valve appears in the brewery.

In his Chapter 5 Moubray looks in depth at failure consequences. Here he considers risk evaluation strategies and looks at operational consequences. His guideline is:

For failure modes with operational consequences, a proactive task is worth doing if, over a period of time, it costs less than the operational consequences, plus the cost of repairing the failure that it is meant to prevent.

Moubray also goes in depth into non-operational consequences and hidden failures. A hidden failure being one that increases the probability of a multiple failure. I have commented on this in the previous section.

Moving on to pro-active maintenance tasks, Moubray goes on to review the failure patterns of Nowlan and Heap and looks in great detail at the potential failure (P-F) curve in Chapter 7 of his work. He defines the working groups/resources required to develop the maintenance system, all leading up to his RCM2 decision diagram introduced in Chapter 10. After the detailed description of this, his following Chapters describe the application and implementation of RCM.

My approach respects most of the principles of RCM as proposed by Nowlan & Heap and reinforced by Moubray. The biggest difference in my approach to that of classical RCM is to include the hierarchical structure as the guide for analyzing the equipment, and to use the MSI concept to define the depth of analysis. The hierarchical structure is then duplicated in the CMMS (which they did not have then) and allows us to easily copy and paste common sections or assemblies, allowing the Planned Maintenance system to be built efficiently.

Using the hierarchy and MSI approach makes the definition of functions at asset and assembly level redundant. Instead, we will look directly at each MSI and how to apply the necessary maintenance strategies to prevent its failure.

4.2 RCM 3

Whereas RCM 2 looks at the consequence of a failure, RCM 3 looks at the risk of a failure occurring (Basson, 2019). RCM 3 is a risk-based methodology aligned with International ISO standards (55000 and 31000). The result is a more rigorous way for developing asset care and risk-mitigating strategies for assets, and has a clear value in high risk industries such as aerospace, nuclear and petrochemical.

Since the release of the ISO 31000 and ISO 55000 Standards for Risk Management and Asset Management respectively RCM3 has been developed as a risk-based RCM methodology that places managing the risk and reliability of physical assets mainstream with other business management systems in an organization. RCM3 exceeds the requirements of the SAEJA 1011 Standard and other ISO Standards.

RCM3 features the following principles:

- Management of physical and economic risks.
- Updated approach for testing and managing of protective systems.
- Based on the requirements of Industry 4.0 Integrates reliability & risk management with organizational management systems.
- Aligned and integrated with International ISO Standards for Physical Asset Management and Risk Management (ISO 55000 & ISO 31000).

Even more than RCM2, RCM 3 being ISO complaint will generate a mass of interlinked documentation, but still has no answer to actually writing effective Planned Maintenance schedules and building them in a hierarchy that allows them to be easily shared across the organisation, to minimize the huge wasted effort inherent in FMEA processes.

What RCM 3 will allow you to do is analyze and document the risk of every failure occurring, and the reduction in that risk from your maintenance plan, which is indeed important in building aircraft, oil platforms and power stations.

However, for the beverage industry we should stick to the basics and build a Common Sense Planned Maintenance system.

Clifford Jones

CHAPTER FIVE: THE NEED FOR THE HACS

The official history of the HEINEKEN Company is well documented on its own websites: https://www.HEINEKEN.com

I have travelled and worked across Africa, the Middle East, Europe and Asia, and have worked for several of the multinational brewers. It was always very clear to me that HEINEKEN is by far the most professional of the major brewers, and it was not by accident that I joined Heineken in 2004 or that I have stayed with them for so long.

They are professional in terms of their product quality and marketing and also in terms of policies, procedures and governance. As employees we are all reminded constantly that we will take no shortcuts in quality, we will never damage the environment, we will always be fair and just leaders who respect diversity and care for the safety of all stakeholders. These things are embedded in the culture of the company, because they are reinforced by senior Manager's every day. As we learn to live in a fully connected and transparent world, with every action we take open to comment and criticism in social media, HEINEKEN has fine-tuned its strategies in sustainability and inclusiveness and has not been found wanting.

5.1 HEINEKEN'S SUPPLY CHAIN

One of the things that HEINEKEN does very well is that whichever country you are in, when you buy a HEINEKEN beer it always looks and tastes exactly the same. This is achieved via very strict quality control standards in production, very strong governance over materials, processes and procedures, and not least by the fact that every brewery in the world has to send samples every month to a taste-panel in Holland, where these very high standards of consistency are maintained.

For many years the Supply Chain was very quality focused, perhaps over so. When I joined in 2004 the Brewing Technologists pretty much ran the Supply Chain and at that time most improvements and initiatives were quality related.

HEINEKEN's Supply Chain started to implement Total Productive Management (TPM) in the late 2000's, and over the past years has achieved sustained improvements in efficiency, reduction in losses and higher productivity.

TPM is a process of continuous improvement that uses cross functional teams (pillars, Figure 12) to analyze performance, carry out deployments and make lasting improvements using a set methodology (called a pillar route).

A few years ago, I saw a presentation at the Global Supply Chain conference showing the trend in operating efficiency, losses, cost of production and productivity over the past 10 years and the results achieved are very impressive.

Of course, HEINEKEN is not alone in this respect, and other major brewers and beverage and FMCG companies use TPM, TQM (Total Quality Management) or six sigma methodologies to achieve continuous improvement.

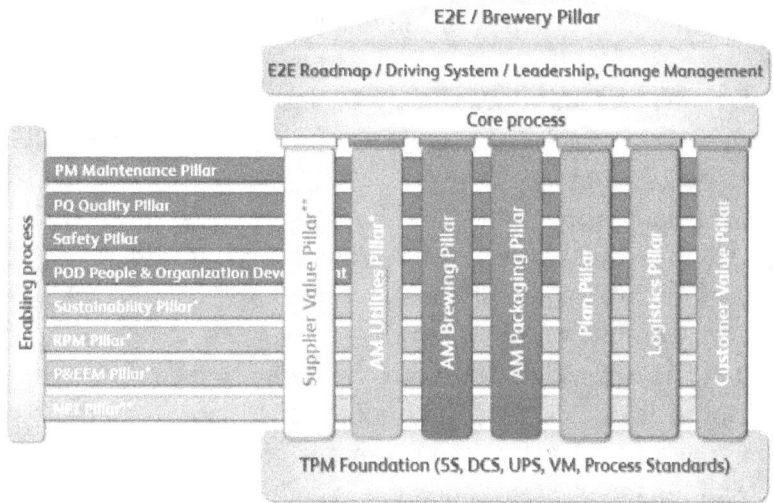

Figure 12, TPM pillar structure 2017, from TPM presentation

Overall HEINEKEN is highly competitive and very successful at reducing losses, improving efficiency and improving productivity in the Supply Chain, all achieved by a combination of thorough TPM application and the sharing of best practices across all of the breweries. Our Supply Chain is extremely disciplined, we have clear KPIs, and we allocate resources to make sure that those KPIs are achieved.

By comparison, the progress in Planned Maintenance over the past 20 years in HEINEKEN has been somewhat fragmented and sporadic.

5.2 CLIFF: WE NEED TO FIX PLANNED MAINTENANCE!

Having worked for HEINEKEN in Egypt, Nigeria and Vietnam, in early 2013 I was promoted to the Asia Pacific Regional office in Singapore as Area Supply Chain Director. All the breweries in the Asia Pacific region (about 22) reported to the Senior Regional Supply Chain Director, who was my boss.

My job was to be his number 2, we refer to it as a "span-breaker role". He managed the bigger breweries in the region, such as Vietnam, Indonesia, China, Singapore, India, and I got to look after the smaller ones, including Thailand, Sri Lanka, Cambodia (then it was still small), Mongolia, Laos, Papua New Guinea, Solomon Islands, New Zealand, Malaysia, New Caledonia and Timor Leste.

At this time there was an Engineer in the Singapore regional team who led the implementation of the Maximo CMMS (Computerized Maintenance Management System, Chapter 11) in the breweries. Initially it was used mostly to control spare parts, but it was an important foundation for our later work.

In early 2017 Jan Paul Boon took over as the Senior Regional Supply Chain Director for Asia Pacific.

In his previous roles Jan Paul had been very much involved in the development of Sahara, a project which looked at the capabilities of each brewery in terms of Engineering maintenance and followed the TPM route to build a Planned Maintenance system.

When Jan Paul moved to Singapore he took some time to evaluate the Asia Pacific Supply Chain, including some joint visits that we made to a few of the breweries together. I think that he realized that my knowledge of RCM and Planned Maintenance systems and how to apply them was quite extensive. So it was that in the middle of 2017 we had a strategy meeting where he informed me that as we were performing poorly at Planned Maintenance in APAC (Asia Pacific), we needed to "Fix Planned Maintenance" in the region. This was something I really wanted to do, and I am sure Jan Paul saw that, but actually getting the green light to start on such a project was quite disconcerting.

The Sahara project had made quite some improvements in basic maintenance standards in Africa, but in APAC we still had many breweries where maintenance was extremely poor and production was restricted by frequent breakdowns with equipment in poor condition, as shown in Figure 13. Some breweries had unplanned downtime as high as 40% (see Figure 25), meaning that Technicians were only reacting to breakdowns (Corrective Maintenance) and there was very little being done except to repair those breakdown's and hope that they did not recur.

Smith (Smith, 1993, p. 3) lists ten common maintenance problems as:

1. Insufficient proactive maintenance
2. Frequent problem repetition
3. Erroneous maintenance work
4. Maintenance practices not institutionalized.
5. Unnecessary and conservative Planned Maintenance
6. Unclear rationale for Planned Maintenance actions
7. Maintenance program lacks visibility.
8. Blind acceptance of OEM inputs
9. Planned Maintenance variability between similar units (See Chapter 12)
10. Lack of Condition monitoring/Predictive Maintenance applications.

I think we experienced all of the above problems in one way or another.

There was a small team (about 5 people) in the Global Head Office in Holland responsible for improving maintenance capability, who were focused on rolling out the Sahara project, which is described later in this chapter. In APAC the TPM Planned Maintenance Pillar route was not bringing the right focus to upgrade Planned Maintenance effectively (Figure 13).

Figure 13: Examples of maintenance failures, from author's maintenance training material.

In our strategy meeting we talked about the Sahara project and what that had achieved in building capabilities, and we talked about the HPO model (High Performing Organisation, Chapter 17), which is a structured way of effectively making improvements, whereby you take a holistic approach and look at Processes and Tasks, Organisational Structure, Reward and Recognition, Governance, Information Systems and People and Competencies and ensure that you address each of these areas to support the change you are trying to make.

Jan Paul was a strong supporter of the Sahara project and of the HPO approach, for good reasons.

We also talked about the "Supportability" project, which was a prior attempt to implement RCM based Planned Maintenance that had been launched by the Global Maintenance team and had engaged some key suppliers to carry out Failure Mode and Effect Analysis for some of the machines that they supplied to us. The idea was that these FMEA tables would later be converted into Planned Maintenance schedules. Unfortunately, due in part to an over focus on completing FMEA templates, the project did not deliver any Planned Maintenance schedules, it is described in more detail in the following section.

In order to "fix Planned Maintenance" in a region where we had so many breweries how could we conduct all of the required FMEA's and make all of the Planned Maintenance schedules for so many machines? Just our breweries in Vietnam have over 2000 separate production machines, so for the whole APAC region the total is probably about 7000, excluding India.

I knew that car manufacturer's produce good, Planned Maintenance schedules for their car's, and that they see the benefit of this because they have many car models that are identical (sometimes millions), and because having cars break down was never to their benefit.

Our equipment manufacturer's produce hundreds of (nearly) identical machines that they sell to brewers and soft drinks companies around the world, the packaging lines used in both of these industries are almost the same, both are filling a carbonated beverage into a container. Their business model is to fly their Technicians around the world to overhaul their machines and install a significant batch of previously ordered overhaul parts. There is no benefit to them in having good maintenance schedules for their customers to use and apply their own Planned Maintenance procedures, quite the opposite in fact. Over the life of the assets, they derive more income from the sale of spares than from the initial sale of the equipment, so why would they challenge a successful business model, even though it is based on the strategies that were abandoned in other industries in the 1970's?

I wanted to follow the auto industry (or even the aviation industry), I wanted to identify the machines we had that were the same, and then produce a set of Planned Maintenance schedules for these machines that could then be easily copied to the other identical/similar machines. This would cut out the unnecessary development of schedules at many different breweries for the same machine, which was an effect of the current TPM approach. Developing Planned Maintenance schedules for every machine at each location in isolation was a massive waste of resources, but it is embedded in the approach of Moubray and others in RCM.

Perhaps at this point most professional Engineers would make a detailed overall plan and timeline, list all the resources and time required, and reach the conclusion that it was a several year project requiring tens of thousands of man hours, calculate a cost for that and then become disillusioned and give up.

But I am a great believer in eating the elephant one leg at a time. (Q: How do you eat an elephant? A: One leg at a time).

I was challenged (by various parties) several times in the first two years of developing the HACS to make an overall rollout plan, to build a governance model, to consider how we will monitor implementation and to utilize the Supportability project work, but I refused to do any of that in the beginning (It comes later), I wanted to concentrate only on eating the first leg of the elephant. If I had looked up and saw the whole size of the project I would probably have given up.

5.3 THE SCALE PROJECT

When I joined HEINEKEN Vietnam in 2009, the brewery in Ho Chi Minh City had one canning line from Sidel/Gebo (an Italian packaging machinery supplier), and we needed to order a second canning line to meet growing demand. At the same time, our Global Projects and Engineering (GP&E) team was commencing the SCALE EQUIPMENT project.

GP&E are the central team designing and installing our major equipment expansions all over the world, and in a company as large as ours they buy new packaging lines and new brewhouses on a weekly basis. What they had come up with was a catalogue of standardized equipment solutions, so that if you needed a large canning line, you should order the same one from the same supplier, and likewise for brewhouses, fermentation tanks etc. Every brewery thinks that they have unique requirements and preferred suppliers, but the SCALE project went a long way to standardizing the equipment we buy and was key to the early success of the HACS.

For large canning lines, the SCALE EQUIPMENT solution was a canning line of 90 000 cans per hour from Sidel/Gebo-Cermex (one company but sometimes divided as either Sidel or Gebo-Cermex but originally was Simonazzi). As we already had one such line at Hoc Mon, I was very much in favor to have the 2nd line from the same supplier, as this greatly reduces the requirements for spare parts stock holding and for training. In my time in Vietnam, I made sure that we installed the same SCALE canning lines from Sidel/Gebo in our breweries in Da Nang, Quang Nam and Hanoi. Because of rapid expansion at the time (In my four years in Vietnam we increased capacity from 3.5mln to 7mln Hls pa), Vietnam became the role model for the SCALE program.

The benefits across Heineken Vietnam in terms of reduced training and reduced spares holding became quite significant.

Figure 14: SCALE canning line in Cambodia built by Sidel/Gebo, from author's own collection.

Returning to the 2017 meeting with Jan Paul, by then we had 8 SCALE canning lines in the region, all identical lines from Sidel/Gebo, split between the breweries in Vietnam and the brewery in Cambodia (Figure 14), with one more canning line in Singapore that was almost the same but has a slightly smaller capacity.

I did not know how long it would take, or what resources we would need to "Fix Planned Maintenance" in the region, but I had an idea where to start. I figured that if we could start by developing the Planned Maintenance schedules for one Sidel/Gebo SCALE canning line, this could then be quickly copied to the other SCALE canning lines. Jan Paul agreed that this should be the first step (which turned out to take about one year) but we would have to do it using existing resources, trainees and interns, without any external consultants or advisors. This turned out to be significant advantage, if we had used external consultants, we probably would not have achieved what we did.

I did not know if we could "Fix Planned Maintenance", I only knew that I could probably get together enough students and Engineers to make the Planned Maintenance schedules for the SCALE canning line, and I hoped that the way to achieve the next steps would become visible later on.

Actually, this worked quite well, because much of what we learned in the first year, especially regarding using a hierarchical standardized structure, enabled us to accelerate the Planned Maintenance schedule development in the following years.

As the HACS development progressed, Jan Paul remained a staunch supporter who allowed me to balance my other duties with developing the HACS system, and to assign a disproportionate number of our trainees to the project. Without his support we could not have developed the HACS, and today he remains one of its strongest advocates, encouraging Heineken teams in many breweries to implement the HACS.

5.4 THE SAHARA PROJECT

The Sahara project aimed to improve maintenance practices and reduce unplanned downtime in the less advanced breweries. Whilst it did not lead to the development of a Planned Maintenance system, it did put a lot of essential basics in place that the HACS was later able to build on.

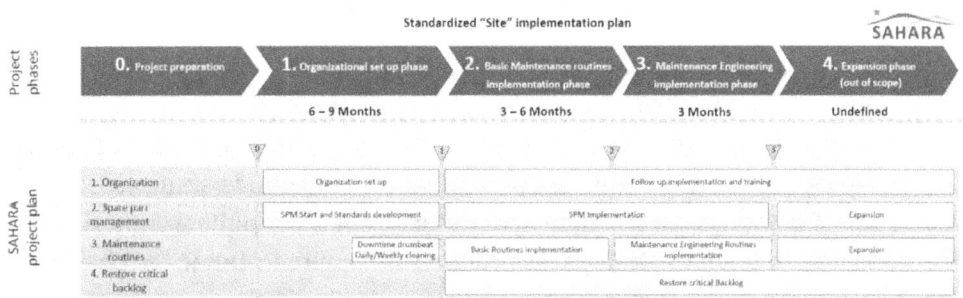

Figure 15: Sahara implementation plan from Sahara roll out presentation.

The Sahara project commenced in the Africa region in 2011 and from 2012 to 2014 was rolled out in 14 of the breweries in Africa. It was well designed and for good reasons focused on the organisational set-up (right structure with required capabilities), Spare Parts Management (SPM) and then the implementation of basic maintenance routines (Figure 15). For an organization with a low maintenance capability this is a good way to start.

Each step had a gate review process to make sure that the building blocks were in place before moving on to the next step. For example, many breweries did not have a well-organized spare parts store or a well-controlled ordering process for spare parts, so it was sensible to put these steps in place first.

Each brewery was regularly surveyed to establish its progress in implementing the various steps.

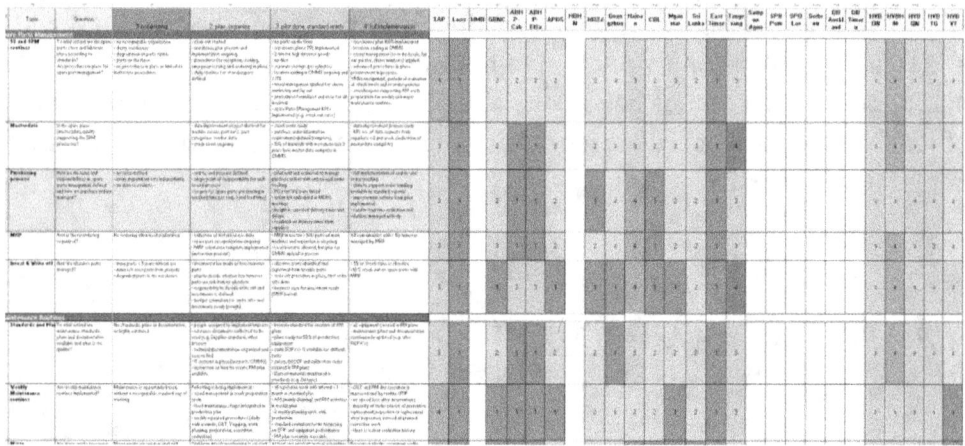

Figure 16: Sahara status map for Asia Pacific, 2017, from author's own files.

In Asia in 2016 I had become involved in supporting the Global Maintenance Team to implement Sahara in the APAC region, and in Figure 16 I have shown the status map for the APAC breweries at that time.

I assisted the global team to hold workshops in the region to implement the Sahara process.

I added my own perspective to the workshops, how I would want to see spare parts organized, examples of poor maintenance that I had seen, and I tried to make the workshops more practical and hands on. I explained the use of gauges in preference to measuring devices and started to introduce some structure into developing Planned Maintenance schedules. I even brought a motorcycle into the workshop to have the delegates practice writing maintenance schedules in a hands-on way.

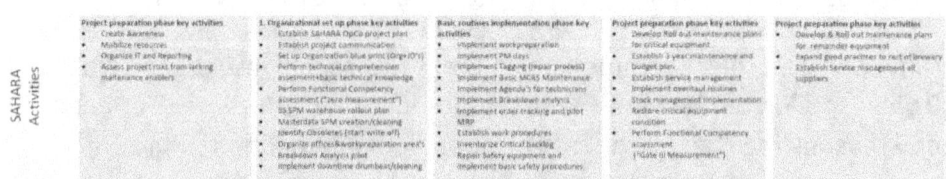

Figure 17: Sahara activities from Sahara roll out presentation.

However, in the breakdown of Sahara activities (Figure 17) there was very little attention paid as to how to write the Planned Maintenance schedules. In fact, here are the only activities from the Figure:

- *Develop roll-out maintenance plans for critical equipment.*
- *Implement overhaul routines.*
- *Develop and roll out maintenance plans for the remainder of the equipment.*

So, although the Sahara project made improvements in the organization and set up of the maintenance function, it did not give sufficient emphasis on developing Planned Maintenance schedules.

5.5 THE TPM PM PILLAR ROUTE

In our TPM methodology, we had a Planned Maintenance Pillar route that was supposed to guide the Planned Maintenance teams:

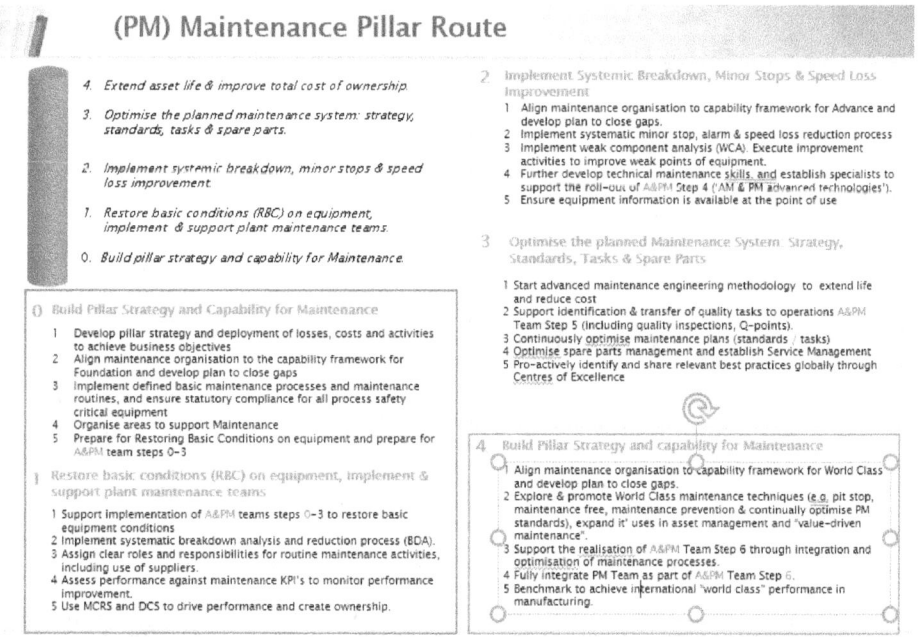

Figure 18: Planned Maintenance Pillar route 2017, from TPM Pillar route presentation.

As shown in Figure 18, the Planned Maintenance Pillar route was very similar to the Sahara implementation plan. It also (correctly) focuses on organization, capability, assigning roles and use of the MCRS (Management Control and Reporting System).

In terms of actually developing Planned Maintenance schedules, it only mentions:

- Step 0.3: Implement defined basic maintenance processes and maintenance routines.
- Step 3.3: Continuously optimize maintenance plans.

So out of 25 activities, only 2 relate to developing Planned Maintenance schedules, and it gives no indication of how the schedules will be written.

But in the TPM system there is a "team route" that explains how to write maintenance standards. The idea is that the Planned Maintenance Pillar will launch a team to write Planned Maintenance schedules, and this team should follow the route in Figure 19.

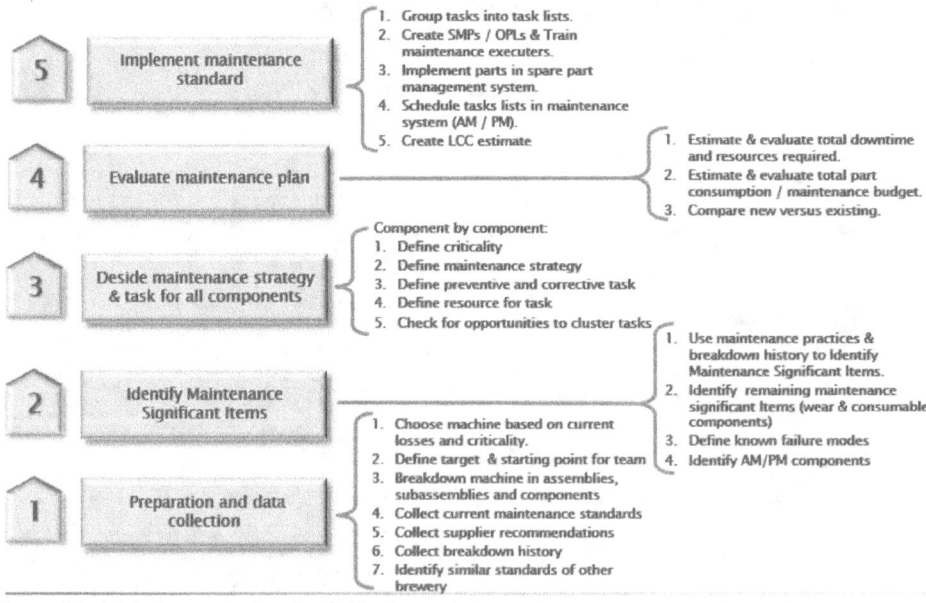

Figure 19: Maintenance standards Team route

If we examine Figure 19 further, the steps given are mostly (but not completely) in line with ways of working that we developed for the HACS.

But in order to carry out these steps, the team members need training in RCM, and they need some standardized structure to work in, such as a framework hierarchy and a template for FMEA.

The result was that many teams following this route completed maintenance schedules for a machine in an unstructured or superficial way: there was no thorough and systematic development of Planned Maintenance schedules.

Another issue with the Maintenance Standards Team Route, is that it encourages teams to separately develop Planned Maintenance schedules for each and every plant item in every brewery, without much consideration of what can be shared and copied, either at machine level or at component level.

The final problem is the availability of the Engineer or Technicians to write the maintenance standards. Only those with some training or experience in Engineering can write good maintenance standards. But all of our Engineers and Technicians are under great pressure every day to attend meetings and "firefight" many daily problems. To sit down and write Planned Maintenance schedules takes a lot of discipline, and human nature being what it is, we will attend to the more urgent issues first and then frequently find that there is no time remaining to allocate to writing Planned Maintenance schedules.

In summary, Sahara and the TPM Pillar route greatly helped many of our breweries to put basic maintenance structures in place, but little was achieved globally in terms of actually writing good Planned Maintenance Schedules or in moving away from an annual overhaul philosophy to an RCM2 philosophy.

Fortunately, since the HACS has become the way of working, the TPM methodology has been updated and there is now a Planned Maintenance "playbook" that describes how to implement Planned Maintenance, based on implementing the HACS.

5.6 THE SUPPORTABILITY PROJECT

This project is a very important example of an unsuccessful approach to implementing RCM based Planned Maintenance. The project followed very well the principles of FMEA (Failure Mode Effect Analysis) and building of RCM worksheets exactly as described in the works of Moubray (Moubray, 1991, p. 209), Regan (Regan, 2012, p. 76) and others.

It illustrates well the unnecessary work required when a rigid FMEA approach is followed, and it is to prevent such a wasteful and counterproductive approach that I have written this book.

The project produced about 60 excel files for various Krones and KHS packaging machines.

The first 20 columns of each file, shown in Figure 20, is a Bill of Materials (BOM) that the supplier completed for the machine.

The suppliers have listed every component in terms of item number, ID number, description, it's expected lifetime, if it should be kept at the brewery as a spare, if it is in contact with food as well as it's dimensions and material of construction. The supplier has also specified which assembly it is part of, as well as the drawing number and several other classifications, including whether it is an MSI (Maintenance Significant Item, see Chapter 8).

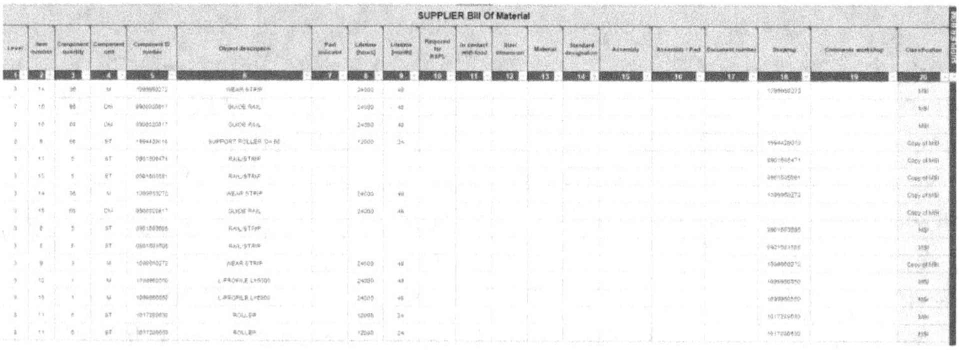

Figure 20: Bill of Materials section of Supportability project.

The next 9 columns in Figure 21, attempt to identify the criticality of each component, again defining assembly and sub assembly and component number, and then assigning a criticality (A, B or C) in terms of whether the machine capacity is reduced by a functional failure and by what percentage it would be reduced.

				CRITICALITY OUTCOME OF MSI's					
MSI No.	Maintenance Significant Item (MSI)	Sub Assembly	Assembly	Component ID number	Maintenance scenario	Criticality [A,B,C]	Impact of functional failure		Remaining machine capacity due to functional failure [%]
22	23	24	25	26	27	28	29		30
10	WEAR STRIP	Conveyor	conveyor	1099950272	Optimised	B	Broken bottles, bottles fall off belt.		0%
11	GUIDE RAIL	Conveyor	conveyor	0900020617	Optimised	B	Stop the chain. Broken bottles		0%
11	GUIDE RAIL	Conveyor	conveyor	0900020817	Optimised	B	Stop the chain. Broken bottles		0%
6		Conveyor	conveyor, air conveyor						
7		Conveyor	conveyor, air conveyor						
8		Conveyor	conveyor, air conveyor						
10		Conveyor	conveyor, air conveyor						
11		Conveyor	conveyor, air conveyor						
12	RAIL/STRIP	Drive	conveyor	0901603505	Optimised	B	Broken bottles, bottles fall off belt.		0%
12	RAIL/STRIP	Drive	conveyor	0901603505	Optimised	B	Broken bottles, bottles fall off belt.		0%
10		Conveyor	conveyor, air conveyor						
12	L-PROFILE L=6000	Drive	conveyor	1099960550	Optimised	B	Broken bottles, bottles fall off belt.		0%
12	L-PROFILE L=6000	Drive	conveyor	1099960550	Optimised	B	Broken bottles, bottles fall off belt.		0%
14	ROLLER	Drive	conveyor	1017200630	Optimised	C	No infeed, no filling of pasteurizer		0%
14	ROLLER	Drive	conveyor	1017200630	Optimised	C	No infeed, no filling of pasteurizer		0%
14		Drive	conveyor, air conveyor						
15	FLANGE BEARING PCST 40 AH23	Drive	conveyor	0404000937	Optimised	C	No infeed, shaft not running		0%

Figure 21: Criticality assessment of Supportability project.

The next 16 columns constitute the Failure Mode and Effect analysis (FMEA), shown in Figure 22. (In Chapter 9 I describe the theory and practice of FMEA).

In this section the suppliers identified the component function, the functional failure, the failure mode and the failure cause, which is all standard for an FMEA. The supplier then identified the statistical distribution of the failure, if it is a hidden failure and the mean time between failures and the probability of failure. Finally, we analyse the failure consequences in terms of safety, environment, availability, cost and quality.

Figure 22: Failure mode and effect analysis of Supportability project.

In Figure 23 we now get to the maintenance strategy and ask again if the spare part should be at the brewery. The supplier then defines the maintenance task, if an SOP is required, CILT is required and the task frequency.

But note that the maintenance task is not described in any detail.

MAINTENANCE STRATEGY				DEFINITION MAINTENANCE TASK							
Maintenance strategy	Spare part brewery [Y/N]	Spare part supplier [Y/N]	Maintenance task	SOP/OPL/S MI required [Y]	SOP/OPL/SMI number	CILT required [Y]	Maintenance task clustering	Additional Information	Task Frequency [Weeks]	AM / PM
49	50	51	53	54	55	56	57	58	59	60
Condition Based (Work after inspection)	Y		Exchange after inspection					Normal mechanic work		PM
Condition Based (Inspection)	Y		Visual inspection of guide rail	Y			A	Normal mechanic work	24	AM
Condition Based (Work after inspection)	Y		Exchange guide rail					Normal mechanic work		AM
Condition Based (Inspection)	N		Visual inspection	Y			A	Normal mechanic work	26	PM
Condition Based (Work after inspection)	N		Exchange after inspection					Normal mechanic work		PM
Condition Based (Inspection)	N		Visual inspection	Y			A	Normal mechanic work	26	PM
Condition Based (Work after inspection)	N		Exchange after inspection					Normal mechanic work		PM
Condition Based (Inspection)	N		Visual inspection	Y			A	Normal mechanic work	26	PM
Condition Based (Work after inspection)	N		Exchange the roller					Normal mechanic work		PM
Condition Based (Inspection)	Y		Visual inspection	Y			A	Normal mechanic work	26	AM
Condition Based (Work after										

Figure 23: Maintenance strategy and maintenance task of Supportability project.

The last section (Figure 24) the supplier reclassifies the consequences after mitigating action, and assesses the impact on operations and resources, in terms of the cost of the part, the hours required to do the maintenance, if maintenance can be done during operation and the tools needed etc.

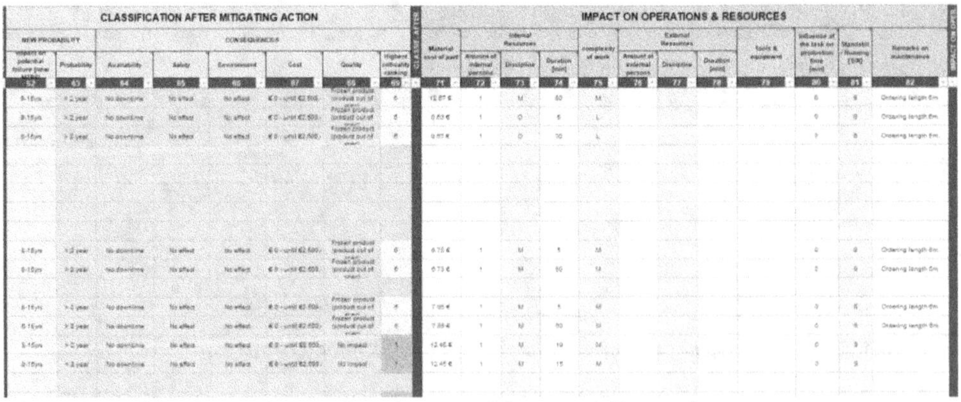

Figure 24: Mitigation and operations impact of Supportability project.

58

After examining these files, it became clear that in the true spirit of RCM2 theory, all of the data for a detailed FMEA was collected, it is a sound and detailed approach to FMEA. It follows very well the principles and if we were an airline, NASA, petrochemical company or if we ran a nuclear power station this might be a very good basis on which to build a robust Planned Maintenance system.

But we are not, we are a brewing company, and such an exhaustive approach is not needed. The difference between the Supportability approach and the HACS approach are illustrated in the following examination of the Supportability file of one machine compared to the HACS file for the same machine.:

- Each Supportability file starts with a complete Bill of Materials. For example, in the file for the Krones tunnel Pasteurizer there are over 7000 line items listed, whereas in the HACS file for the same Krones tunnel pasteurizer we have about 300 items listed. In the HACS we don't list the full BOM, only the MSIs. So about 6700 items in the Supportability file are not needed.
- Of the 7000 items listed by Krones for this machine, 577 are defined as MSIs. In the HACS file for the same machine, we have 82 MSIs. There are 495 MSIs that are not needed, they are not really MSIs.
- The number of maintenance tasks proposed in the Supportability file for the pasteurizer by Krones is 577 (exactly one task for each MSI), but in the HACS for this machine we have only about 120 Planned Maintenance tasks for the 82 MSIs defined.
- Almost all of the Planned Maintenance tasks in the Krones Pasteurizer Supportability file are either condition-based inspection or run to failure. Only 3 times do we see a Time-Based replacement and 1 modification is included. The maintenance strategy is not balanced.
- In the HACS file for the same machine there are 28 Time-Based replacements, as well as a good number of lubrication and calibration tasks, which are missing in the Supportability file.
- The maintenance tasks are not clearly defined. For example, "Visual Inspection of rail". Every inspection has to have a clear standard and has to describe how the inspection is carried out.
- SOPs that are mentioned as required were not provided, presumably this would be a next stage from the suppliers.

Adopting this system would require a massive increase in Planned Maintenance resources as so many inspection tasks would require many Technicians to carry them out. As you will see in the following sections, this illustrates again that our maintenance philosophy in our industry is far behind the aircraft industry.

My conclusion was that the Supportability work was not of any use in building an effective Planned Maintenance system, and that I needed to figure out myself how to do it in an efficient and productive way.

In her book Regan frequently mentions the importance of a Facilitated Working Group approach (Regan, 2012, p. 31), and a good facilitator might have prevented the entire BOM being listed and some of the unnecessary work described above. But in Regan's approach there is still no attention paid to the hierarchical structure (to support roll out to other sites and other machines) and no mention of how to write the all-important maintenance schedules, Job Plans and SOPs

The point to remember, which is well illustrated by the Supportability project, is that we need to develop practical solutions to building a Planned Maintenance system in our industry, and not invest scarce resources in a complex and detailed FMEA that adds little value.

5.7 REASONS FOR RCM IMPLEMENTATION FAILURES

Neil Bloom in his book Reliability Centered Maintenance (Bloom, 2006, p. 15) cautions against the use of consultants to implement RCM, as most have never actually personally implemented an effective RCM based Planned Maintenance program, and they often end up being the only ones fully understanding their process, resulting in a vacuum when they leave. As recommended by Bloom, consultants should only ever be brought in as a temporary augmentation to assist in establishing a program that remains under your own control and direction.

Bloom is one of the few authors with many years of experience in actually developing implementing RCM programs himself, mostly in the nuclear power industry.

Bloom mentions several potential reasons for the failure of an RCM program:

LOSS OF IN-HOUSE CONTROL:
Farming out the RCM analysis to a third party is a major pitfall, and the result of the supportability project illustrates this well. It is essential to maintain control

over the criteria and parameters, to make sure that a proper hierarchical structure is followed, that only likely failures are analyzed, that the correct maintenance strategy is chosen, that Planned Maintenance schedules are copied when they already exist elsewhere for the same component, that Job Plans have clear standards etc.. and above all that the Planned Maintenance program is appropriate for the industry it is being developed for.

INCORRECT PEOPLE CARRYING OUT THE ANALYSIS:

Failure analysis and construction of the Planned Maintenance system must involve the right mix of in-house personnel from the Engineering and Operations teams. Firstly, this ensures that local knowledge is captured, secondly that there is buy-in and ownership of the Planned Maintenance program that is produced.

One of the challenges of the HACS is that because we build a set of Planned Maintenance schedules for a SCALE machine, we expect it to be applied in another brewery exactly as-is, if they have the same machine. This means that the staff at that brewery have to reach the point where they accept and trust the Planned Maintenance system. One way we supported this was to have as many Engineers as possible from different breweries working on the Planned Maintenance schedules of different assets in the HACS. Getting buy-in does remain a challenge as the HACS is rolling out globally into other regions of the world. Where the brewery does not have SCALE equipment then the local brewery has to modify the HACS for their machine, so there the ownership is better assured.

But after all, has been said, if you take your car for a service you generally trust that the service schedule of the manufacturer has been properly designed and you are happy for the workshop to follow that schedule. This is the level of trust we need to reach for the HACS, and we are making great progress.

UNNECESSARY ADMINISTRATIVE BURDENS:

Too many steering groups, focus groups, committees and facilitator training programs are all cited by Bloom as reasons for RCM implementation failure. To this we should include endless FMEA: One consultant that I know of carried out FMEA analysis and managed to find an average of 10 different failure modes for each functional failure, which is an outstanding amount but really not necessary in our industry, and the sort of unnecessary administrative activity that will cause the program to fail.

FUNDAMENTAL RCM CONCEPTS NOT UNDERSTOOD:

The most fundamental concept in RCM is that the assumption that every component has a certain lifetime after which it must be overhauled, inspected and if necessary, replaced is not true. Yet we still have many Engineers in our company who follow the practice of big annual overhauls and do not understand what was discovered by Nowlan and Heap over 40 years ago.

INSTRUMENTS NOT INCLUDED IN RCM ANALYSIS:

This was an area that Bloom found to be neglected in his experience and he devotes a chapter to RCM for instruments (Bloom, 2006, p. 181). Early in the HACS development process I also realized that instruments played a critical role in the overall Planned Maintenance program, and they are an integral part of the HACS. (Refer to chapter 15 where Calibration is discussed).
Bloom mentions further possibilities including:

- Confusion determining system functions.
- Confusion concerning system boundaries and interfaces.
- Divergent expectations.
- Confusion regarding conventions.
- Misunderstanding of hidden failures and redundancy.
- Misunderstanding of run to failure.
- Inappropriate component classifications.

I think that the last points might slow down the development of a Planned Maintenance program greatly but could still be overcome with the right in-house ownership and control over the process.

Next, we look in detail at the Failure Process and Failure Patterns, as these need to be fully understood before we can start to develop a Common Sense Planned Maintenance system.

CHAPTER SIX: THE FAILURE PROCESS AND FAILURE PATTERNS

In the 1990's when I was working for the South African Breweries (SAB) I attended a workshop on Planned Maintenance, led by Bill Hughes who was then the Senior Engineering Consultant for SAB, a position that he held from 1987 to 1995.

Bill was a down to earth and inspiring presenter. He talked about Planned Maintenance in clear and simple terms. I recall his analysis of the possible reasons for the failure of a conveyor gearbox, and his clear statement that there were only 4 possible failure modes (he used colourful language), and that in this case Failure Mode and Effect Analysis does not need to be a complex and detailed process, we just need to address the obvious failure causes, check the oil and find a way to check for wear on the bearings or gears.

Bill believed in practical and effective solutions in our industry, and the seeds he planted did bear fruit nearly 30 years later when I started building the HACS.

6.1 DOWNTIME & BREAKDOWNS

In the Brewing and Beverage world, and probably in most of manufacturing, we aim for continuous production in order to maximise output from the assets that we have. Our equipment is a considerable investment, tens or hundreds of millions of dollars, and it pays back on that investment only by producing the goods that we sell.

We thus measure and try to control output and especially the non productive time.

When non productive time is due to a product changeover, periodic cleaning or a lack of demand for product we call it planned downtime.

Unplanned downtime can be due to a lack of materials or operators, power supply cut, lack of product from the upstream process, or due to a machine breakdown.

We should be able to minimise or eliminate unplanned downtime due to materials, upstream product etc..with good planning (and perhaps also reducing some planned downtime, for example by having less changeovers).

Most breweries now apply an S&OP process (Sales and Operations Planning) that streamlines communication between the various departments and enhances alignment on the production requirements and co-ordinates end-to-end through Procurement, Warehousing/Logistics, Human Resources and Production.

So with the Plannning issues hopefully eliminated, then the remaining unplanned downtime will normally be due to breakdowns of the equipment.

Within our region there was a huge variation in unplanned downtime:

Region	OPCO name	Unit name	UPD%
APAC	Cambodia	Phnom Penh	9.3
APAC	Indonesia	Sampang Agung	22.8
APAC	Indonesia	Sampang Agung SOFT	4.6
APAC	Indonesia	Tangerang	18.7
APAC	Laos	Vientiane	32.6
APAC	Malaysia	Petaling Jaya	13.5
APAC	Myanmar	Yangon	11.8
APAC	New Caledonia	Noumea	22.6
APAC	New Zealand	Auckland	27.0
APAC	New Zealand	Nelson	10.9
APAC	New Zealand	Timaru	17.2
APAC	Pap. New Guinea	Lae	39.8
APAC	Pap. New Guinea	Port Moresby	41.7
APAC	Philippines	Cabuyao	47.6
APAC	Philippines	El Salvador	33.9
APAC	Singapore	Tuas	37.2
APAC	Solomon Islands	Honiara	27.7
APAC	Sri Lanka	Mawathagama	31.7
APAC	Timor-Leste	Hera	4.1
APAC	Vietnam Brw	Danang City	11.4
APAC	Vietnam Brw	Hanoi	17.6
APAC	Vietnam Brw	Ho Chi Minh City	7.4
APAC	Vietnam Brw	Quang Nam	2.3
APAC	Vietnam Brw	Tien Giang	8.6
APAC	Vietnam Brw	Vung Tau	9.3

Figure 25: Unplanned downtime in the APAC region 2018, author's files.

As shown in Figure 25, unplanned downtime can vary from 2% to almost 50%. Knowing that the majority of this is due to equipment breakdowns, the above shows clearly that there is a vast difference in equipment condition and in maintenance capability across the different breweries.

Some of this difference in unplanned downtime can relate to equipment age, and also to complexity. For example, a canning line will almost always have less unplanned downtime than a returnable bottling line, as the materials are more standardised on a canning line (no broken/chipped bottles), and the processes are simpler (no bottle washer or bottle inspector). Some breweries have a large number of SKU's and thus many changeovers, and this can impact both planned and unplanned downtime, as there are often many small issues to attend to after a changeover.

Sometimes we find that older equipment, if well maintained, will have less unplanned downtime because it is simpler, with less sensors and electronics, and therefore there are less parts to fail.

What is clear from the above is that we need to reduce unplanned downtime in a sustained way, and that means reducing equipment breakdowns.

In our scope (the beverage industry): A breakdown is when a production asset is unable to produce products in specification due to one or more component failures.

So it may be not able to operate, or it may operate with the wrong specification, such as producing damaged cartons or out of specification beer.

6.2 DEFINITION OF A FAILURE

In the above definition of a breakdown I mention component failures as the cause of a breakdown.

Nowlan and Heap in Reliability Centered Maintenance (Nowlan, 1978, p. 18) define that *"a failure is an unsatisfactory condition"*, and that *"a functional failure is the inability of an item to meet a performance standard"*.

Moubray defines a functional failure (Moubray, 1991, p. 47) as the *"inability of any asset to fulfil a function to a standard of performance which is acceptable to the user"*.

In this definition Moubray defines the failure relative to the asset, and as an example describes how a pump, the asset in this case, can fail to fulfill or partially fail to fulfill it's function. I prefer to define the failure relative to the component, as we usually see the asset as the whole machine, and we tend in practice not to use the word "asset" when we are dealing with single components.

In the example of Moubray I would refer to the components that have worn or failed. It can be that the impeller of the pump is worn causing a low flow rate (partial loss of function), or it can be that the impeller is disconnected from the keyway on the shaft, causing a complete loss of function. For me it is not sufficient to just state that the pump has failed.

This is an important difference of view, and it comes from practical experience wherein we always have to inspect, replace or repair the failed component, which we refer to as a Maintenance Significant Item (MSI).

The Maintenance Significant Item (MSI) is a key definition in building a maintenance system that we will discuss further in Chapter 8.

Leaving the MSI definition to one side for now, we can combine the definition of Moubray with my own experience and come to a Manufacturing definition:

A functional failure is when a Maintenance Significant Item is unable to fulfill it's designed function(s) causing the breakdown or reduced performance of a production asset.

A breakdown is when a production asset is unable to produce products in specification due to one or more failures of Maintenance Significant Items.

Reduced performance would cover any area where the asset does not reach it's original designed operation but is still producing. It could be that a safety device is not operable, for example, but the asset is still utilised in a bypass or manual mode.

Going forwards we will use the above definitions.

6.3 THE FAILURE PROCESS:

Usually components don't just suddenly fail instantly and randomly. They do tend to wear over time, and this led to the assumption that *"every component has a certain lifetime after which it must be overhauled, inspected and if necessary replaced"* which was challenged and quantified by the work of Nowlan and Heap.

Nowadays we refer to a failure process, and in Chapter 7 of his book, Moubray shows the P-F curve, and relates this back to the use of Predictive Maintenance, noting well that the point where the failure starts to occur may not be related to age of operation of the asset (more correctly, component).

In my RCM training I use the slide shown in Figure 26:

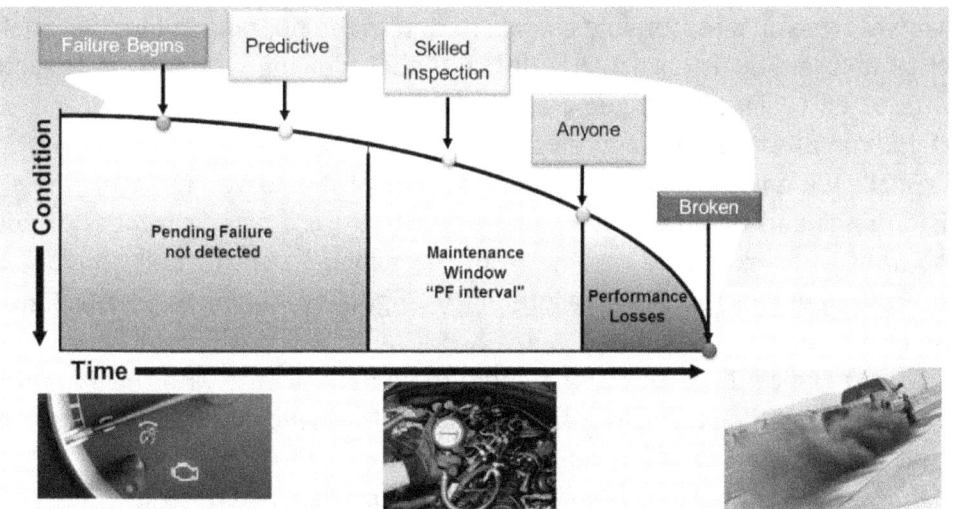

Figure 26: The failure process from author's RCM training materials.

In the first section of the curve, the pending/potential failure is not detected and can only be foreseen by predictive maintenance. Here I use the example of a maintenance interval indicator that is common on many cars today. The maintenance warning light is activated by an algorhthym written by the manufacturer that takes into account your driving style and other key inputs from the cars sensors. If you accelerate hard and brake hard then the light indicating that maintenance is required will come on much sooner than if you drive carefully and sensibly. The maintenance warning light is a good example of a simple Predictive Maintenance device.

Later the engine wears more and we enter the maintenance window, the P-F interval where the pending/potential failure can be detected with skilled inspection. If we consider the wear on the pistons in the engine, then a skilled inspection is to carry out a cylinder head pressure check, which quickly indicates if either the piston rings or cylinder bores are worn and we will need to replace them. It is also possible at this stage to tear down the engine and measure the ovality of the cylinders and wear on the piston rings. So this would be a choice between Condition Monitoring and Inspection, more about that in Chapter 8 on Maintenance Strategies.

The length of the Maintenance window (P-F interval) is very important when we try to carry out inspections. For example, if the window is short, lets say that the engine moves from having to low compression to complete failure in a few weeks, then we can only detect this pending failure if we have very frequent inspections. However, if the window is longer, we can detect with less frequent inspections, such as annual ones.

Whenever you decide on an Inspection task, you have to estimate a frequency that will give you a reasonable chance to detect the pending failure (P-F) before actual failure. This is particularly difficult when dealing with fatigue in aircraft parts, as early signs of fatigue cracks may be very difficult to detect, and early fatigue can propagate to failure very unpredictably.

Finally we enter the performance loss part of the curve. In our example the piston rings and cylinders are so worn that there is a significant loss of power and a lot of black smoke coming from the exhaust. At this point anyone can detect the pending/potential failure of the engine by looking at the black smoke coming out of it.

At first glance this process may seem to be fully aligned with the assumption that *every component has a certain lifetime after which it must be overhauled, inspected and if necessary replaced.*

But we don't know why (and when) the failure began in this example. It could have been that the air filter was not changed and became torn, allowing sand and dirt into the engine. It could have been that the driver accelerates hard and revs the engine when the oil is still cold, causing excess wear.

Actually, the life of a piston engine is very dependent on how it is run-in (how the engine is operated in the first hours of running), and there is good reason for the running-in instructions of the manufacturer when you first use a new car. Technically this relates to the piston rings bedding in and wearing away the very fine machining imperfections left on the surface of the cylinder bore. Usually, engines have an early first oil change, aimed to remove all the small metal particles in the oil coming from the running-in.

Suffice to say that the life of an engine is probably as much dependant on the driver's actions as it is on the operating hours (Something I was in complete denial of as a young man), and probably we can say the same in a brewery. The life of the components in the brewery will also depend on how we operate and maintain the equipment.

If we carry out the correct lubrication and cleaning, we can expect a longer life than if we do not. If we spray everything with high pressure water guns once a week, surely the lifetime is less. In many developing countries where we operate we see a lot of electrical component failures that are caused by fluctuations in the power supply, such as spikes or low voltage.

In summary, so far we know that failure is a process that can only be detected in the P-F interval maintenance window, but that the process is usually not dependant on operating hours or cycles, but often on other factors.

6.4 THE FINDINGS OF NOWLAN AND HEAP IN DEPTH

When I deliver RCM training to students and Engineers, I usually show the failure curves as first presented by Nowlan and Heap but without the percentage occurrence, and then ask my students to estimate the percentages themselves (Figure 27).

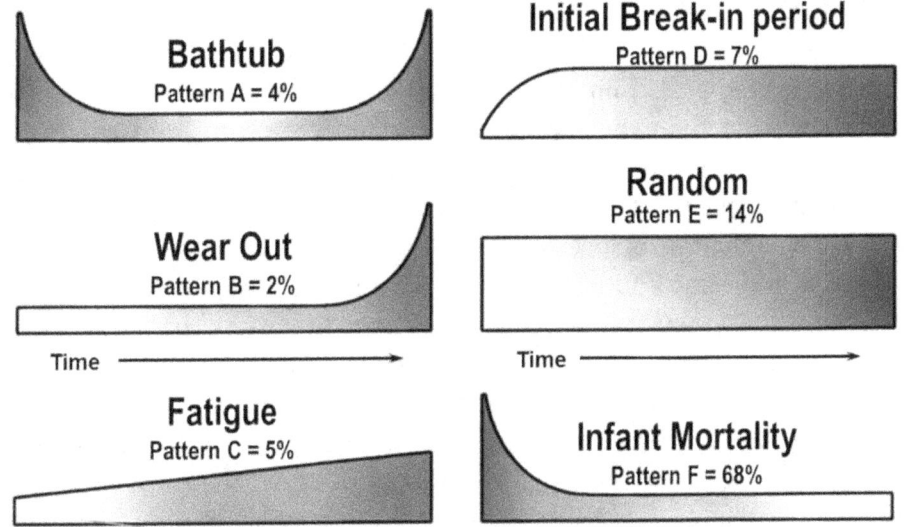

Figure 27: The failure curves of Nowlan and Heap, from the author's RCM training material. Horizontal axis is the hours of operation, vertical axis is the probability of failure.

Most of the answers are very far away from the findings of Nowlan and Heap, and wear-out is always estimated to be much higher than shown above.

Let's examine their findings:

PATTERN A, THE BATHTUB CURVE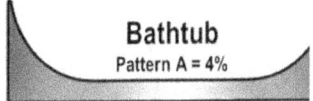

In this failure pattern there is an initial high probability of failure relative to time, known as the Infant Mortality part of the curve, that quickly falls and is followed by a period of constant or very slightly increasing failure probability, which is then followed by a Wear-Out region where the probability of failure again becomes very high.

Where components follow the Bathtub curve failure pattern, then replacement after a certain time of operation, before the failure probability increases, would be a sound maintenance strategy. For this to be effective, you have to know when the probability of failure increases for that component.

Nowlan and Heap found that this failure pattern applied to only 4% of the components that they examined in aircraft.

PATTERN B: THE WEAROUT CURVE

Wear Out
Pattern B = 2%

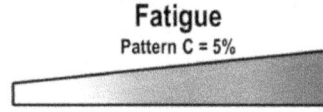

The previous Bathtub curve is a combination of this Wear-Out curve and the Infant Mortality curve. The Wear-Out curve illustrates a period of constant or very slightly increasing failure probability, which is then followed by a Wear-Out region where the probability of failure becomes very high.

Even more-so than the Bathtub curve failure pattern, replacement after a certain time of operation, before the failure probability increases, would be a sound maintenance strategy, and with this curve there is no associated risk of Infant Mortality after replacement.

Nowlan and Heap found that this failure pattern applied to only 2% of the components that they examined in aircraft, and they state that it is characteristic of reciprocating (piston) engines. However, an aircraft engine is itself a huge number of parts, and at the time of their research aircraft piston engines were becoming very complex (for example radial engines of up to 36 cylinders), so I personally suspect that aircraft piston engines could themselves follow many different failure patterns depending on the many individual components within them.

But it is relatively easy to think of components that follow this failure pattern, such as a car tyre. As the tyre wears and becomes thinner, it reaches a point where the remaining tread is very thin, and thus the probability of failure from a sharp object becomes very high (failure resistance has lowered).

PATTERN C: THE FATIGUE CURVE

Fatigue
Pattern C = 5%

This curve shows a gradually increasing failure probability, with no identifiable wear out section. However, do note that the probability of failure is directly proportional to operating hours, and this is the only curve displaying this characteristic.

Nowlan and Heap found that this failure pattern applied to only 5% of the components that they examined in aircraft and stated that the pattern was followed by aircraft turbine engines (Nowlan, 1978, p. 46). Turbine engines are, like piston engines, very complex machines. Their fundamental advantage over a piston engine is that they do not reciprocate, they rotate, so that they do not have the same wear characteristic or inherent vibration as a piston engine. But like the piston engine, they are made up of many thousands of components that each have the potential to fail according to different failure patterns.

Fatigue is an Engineering concept whereby there is the initiation and propagation of cracks in a material due to cyclic loading. It is very important in aircraft design, as there are many different cyclical loads taking place on a structure that is designed to be as light as possible.

On April 28, 1988, Aloha airlines flight 243 operated a Boeing 737-200 that suffered a fatigue fracture that released a very large section of the cabin, killing one stewardess (Figure 28).

Figure 28: Fatigue failure of Aloha airlines Boing 737, from Wikipedia

A major contributing factor to this accident was that the aircraft was "Island hopping", so it incurred many more pressurization cycles each day than most aircraft, in fact 89680 cycles at the time of the failure, which was found to be accelerated by saltwater corrosion.

https://en.wikipedia.org/wiki/Aloha_Airlines_Flight_243

As mentioned by Regan (Regan, 2012, p. 20), Boeing designed structural inspections assuming 1.5 pressurization cycles per flight hour, but the Aloha airlines aircraft were accumulating double that, 3 pressurization cycles per flight hour.

In general, we can expect to see fatigue failures where we have cyclical mechanisms. In the 1980's many of the pasteurizers used in the beverage industry had a walking beam that cycled back and forth, and these beams were so prone to extensive fatigue cracking that this type of machine is no longer manufactured.

PATTERN D: INITIAL BREAK IN PERIOD

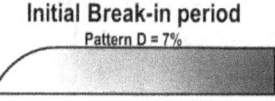

Initial Break-in period
Pattern D = 7%

Here we see a low probability of failure when the component is newly installed, which quickly increases to a constant probability of failure.

Nowlan and Heap found that this failure pattern applied to 7% of the components that they examined in aircraft.

This may be the case with rotating parts, such as a shaft oil seal that will maintain a good seal initially but can rapidly wear after installation to a state of failure.

PATTERN E: RANDOM FAILURE

Random
Pattern E = 14%

Time

Here there is a constant probability of failure regardless of the operating hours.

Nowlan and Heap found that this failure pattern applied to 14% of the components that they examined in aircraft.

For me the only possible failure causes are those that come from the observable laws of physics. I don't accept the existence of a random failure as a fundamental principle. Whilst events can appear to be random, such as a meteorite strike or a weather event, there must be an assignable cause for a failure, even if the cause itself occurred in an apparently random fashion. Components do not randomly fail; they fail because they have entered the final part of the P-F curve. In the next chapter I describe our investigations in this area, but for now, let's consider an electronics part failure:

Generally, we tend to say that electronics failures are random, as we usually don't know what has caused them, so we just replace the failed component. But often a part has overheated because the cooling fan was blocked with dust. Or it can be that a part has failed because of a voltage surge, but we can't normally see that occurrence. Failure can also be due to moisture ingress, or due to electrostatic discharge. If you walk across a nylon carpet and discharge a voltage to your laptop, it is an assignable failure cause. I suspect that fatigue failures are common in electronics, because of the cyclical effect of heating and cooling a circuit board. Think of the soldered joints of the components on a circuit board, and how a minute stress is applied to them every time that the circuit board is heated and expands a very small amount and then shrinks the same small amount as it cools down when it is switched on and off. This stress will eventually cause fractures in the soldered connections, which we likely assign to a random failure, but it was not random. Switching electronic components on and off frequently, allowing them to expand and contract, is surely more likely to lead to a failure than leaving them switched on and in a constant state (Providing they are not consuming too much current). Other failures of electronics can be due to aging, whereby insulation materials break down over time for example, but again this is not random.

Currently I live fairly close to the brewery, and I know that when there is a grid power failure at my home on the weekend, then there will be several electronic failures at the brewery as well. This is because when the power supply is cut and restored there are voltage spikes and these will cause a failure somewhere, which is decidedly not random.

PATTERN F: INFANT MORTALITY

In general use the term 'infant mortality" refers to the death of an infant before his/her first birthday, an event that was much more commonplace before the discovery of vaccines and other modern medical knowledge.

The Infant Mortality failure pattern shows a very high probability of failure immediately after installation of a new component, which then reduces to a steady or slightly increasing probability.

Nowlan and Heap found that this failure pattern applied to an incredible 68% of the components that they examined in aircraft. So, it means 2 out of every 3 failures.

The Infant mortality curve is pretty much the opposite of the Wear-Out curve. The Wear-Out curve means you can replace a part after a certain time of operation. Infant mortality tells you to leave it alone. It is more likely to fail if you replace or overhaul it than if you do not.

From this finding it is now clear that the assumption *that every component has a certain lifetime after which it must be overhauled, inspected and if necessary replaced is* incorrect.

You could reword this to say, 89% of components do not have a failure rate related to operating hours and should not be periodically overhauled (Infant mortality + random + break-in = 89%).

In my younger days we had filament bulbs (as LED's were not yet commonplace) in our light fittings, and these were frequently a good example of Infant Mortality failures. From a box of light bulbs, several would fail instantly when you plugged them in and switched them on. This normally relates to manufacturing defects, as in the case of filament light bulbs, an imperfect vacuum will usually cause this immediate failure.

Infant Mortality can relate to manufacturing defects and also to installation issues (man or method failure).

Counterfeit parts from some countries can contribute significantly to Infant Mortality, as well as a lack of quality control standards in some countries. "*Poorly made in China*", published by Paul Midler (Midler, 2010) gives an excellent description of manufacturing practices in China and goes a long way to explaining how continuous and uncontrolled cost cutting can impact product quality.

As mentioned in Chapter 3, the practice still prevalent in the beverage industry is one of big annual overhauls. As such an overhaul introduces many new components, the majority of which follow an Infant Mortality failure pattern, then we have introduced more failures than if we had never done the overhaul. This is evident in the performance in the period directly following the overhaul. Smith shows (Smith, 1993, p. 21) that in conventional power stations 56% of forced power outages (inability to supply electrical power) occurred within one week of a maintenance overhaul.

In building the HACS I almost completely eliminated the concept of an annual overhaul. As much as possible the maintenance activity is designed to prevent the component failure and is carried out according to the failure pattern of that component.

CHAPTER SEVEN: UPDATING NOWLAN AND HEAP: CHAN HUI LING & FOO PEI SAN

In Chapter 2 I mentioned that I studied many aircraft accident reports as a result of my father having a flying license (initially a private license and later a full Commercial license), and a common feature of many of those accidents was bad decision making on the part of the pilot.

But I also noticed that in those accidents where maintenance was a significant contributor, that there was usually a human element in the maintenance failure.

Very, very, occasionally an aircraft would have maintenance carried out and be handed back to the user with the control wires crossed, having been incorrectly re-assembled. If not spotted before takeoff, this would usually lead to a fatal accident.

On 10 June 1990 a BAC One-Eleven suffered explosive decompression from an improperly installed windscreen panel causing the captain to be partially sucked out of the aircraft (he managed to survive). The root cause of the accident was that the maintenance technician used the incorrect size of securing bolts in the windscreen frame.

https://en.wikipedia.org/wiki/British_Airways_Flight_5390

In my many years in the Beverage Industry I have seen many breakdowns where the 5-Why root cause failure analysis reveals that the failure was caused by incorrect installation (we can call this man or method), either because the technician was not trained, or the procedures used were not correct.

In the aircraft industry there is a very thorough system of training and qualification of Technicians, and very detailed procedures are issued by the manufacturer's regarding how to carry out every maintenance task. But they were probably not perfect in the 1960's so I wondered why Nowlan and Heap do not mention man/method as a potential failure pattern/cause in their paper Reliability Centered Maintenance at any point?

In the Beverage industry it is often the case that technicians are poorly trained, do not have the required competencies or are not familiar with the equipment that they are working on. Much of our equipment has very sparse maintenance instructions, for some machines they are utterly useless, so failures due to incorrect installation will be much more likely for us.

As we developed the HACS and I trained many Engineers on RCM, I began to wonder whether the findings of Nowlan and Heap are really applicable to our industry. Not only that, but when they carried out their study there was not much progress in the field of electronics, and they did not have the alloys and composite materials that we have today. I wondered how much the failure patterns are affected by modern electronics and materials technology?

Finally, breweries don't have the redundant systems that are built into aircraft, we certainly do not have the lightweight construction of aircraft and I doubt that there is such a rigorous design process with critical risk analysis done.

I wanted to verify whether the findings of Nowlan and Heap, which are the basis of the maintenance strategy for countless non-aerospace organizations globally, can really be transposed to non-aerospace industries, and whether they are still valid after more than 40 years.

In 2020 two students from Singapore University of technology who had worked on the HACS development for 6 months as Interns with me, approached me for suggestions for a Capstone project for their Engineering qualification from Newcastle University.

Under my guidance Vivien Chan Hui Ling and Joycelyn Foo Pei San carried out a study of all of the failures that occurred in our breweries in Singapore and Cambodia over a 6-month period with the objective of evaluating if the findings of Nowlan and Heap applied to the Beverage industry today, and in addition I wanted to explore further whether so-called "random" electronic failures were really random.

In the study 42 component failures were examined, here are some of the ones that gave us new insights into the failure mechanisms in our industry.

7.1 SIGHT GLASS FAILURE

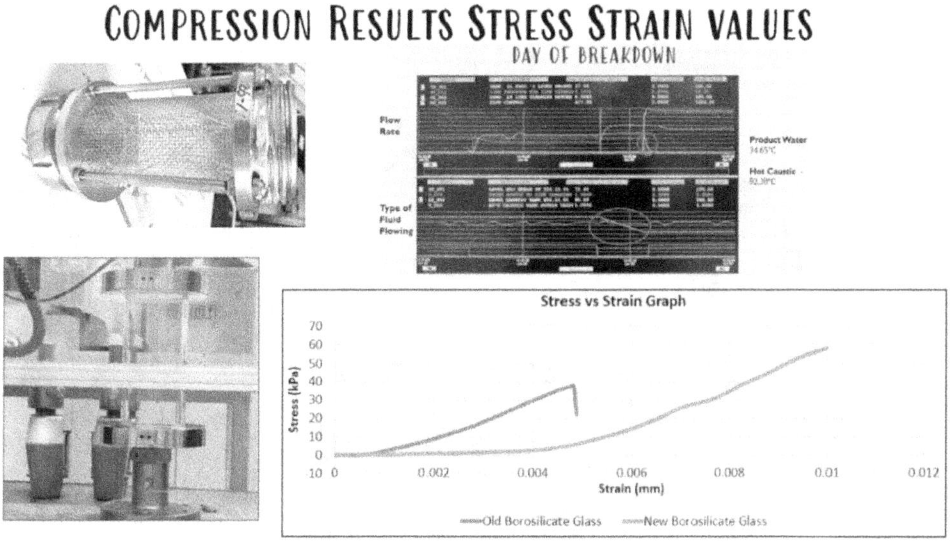

Figure 29: Sight glass failure analysis, from Chan Hui Ling & Foo Pei San Capstone project report.

A sight glass failure on a process pipe caused a significant loss of product. The sightglass is shown top left in Figure 29. The failure related to a pressure fluctuation in the pipe, but the fluctuation should not have been sufficient to cause the sightglass to fail. We know this from the process control data shown top right. We carried out testing of the compression failure point of used (3-5yrs) versus new glasses in a compression test device, shown bottom left, and found that the compression resistance is considerably weakened in a used glass, to below half of that for a new one. This is shown graphically in the bottom right of Figure 29.

It is our hypothesis that the cyclical stress of the hot cleaning process that is carried out periodically, causing expansion and contraction of the fitting holding the sight glass, initiates microscopic fatigue cracks that reduce the pressure resistance of the sightglass.

This is a clear example of a fatigue failure (Pun intentional), but there does not seem to be any non-destructive way of testing for the fatigue, it is hidden in the microscopic structure of the glass.

7.2 DISPLAY PANEL FAILURE

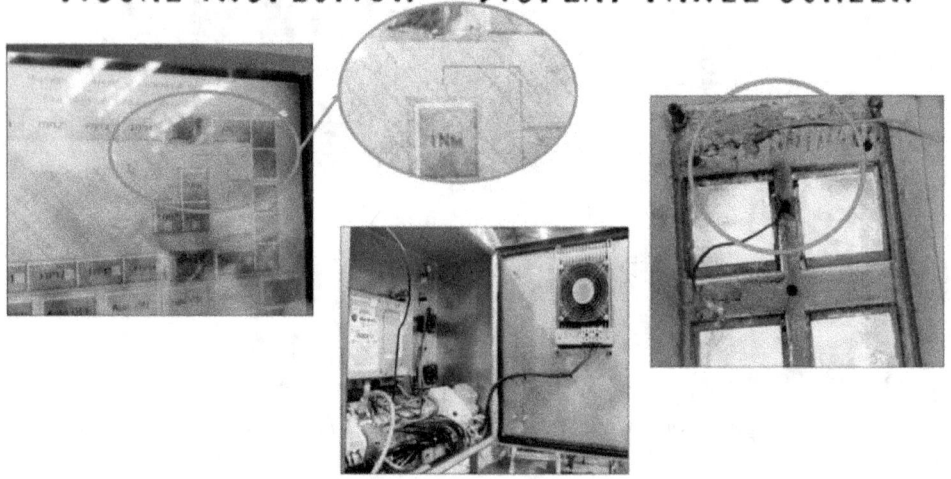

Figure 30: Failure of display panel screen, from Chan Hui Ling & Foo Pei San Capstone project report.

It was discovered that a machine display panel screen, shown top left in Figure 30, had been replaced several times with the same fault of the LCD display clarity deteriorating making it difficult to identify items on the screen.

The cooling fan in the electrical panel housing the display and its integrated computer was found to be not operating, shown in the center of Figure 30. On the right of Figure 30 is the cooling pad that the fan is attached to. It was discovered that the wire to the cooling fan was broken inside the cooling pad, probably due to fatigue from opening and closing of the panel door.

For the LCD display failure, the classification is external causes, as the root cause is positively identified as the cooling fan power cable failure, which is external to the screen.

7.3 VOLTAGE REGULATOR FAILURE

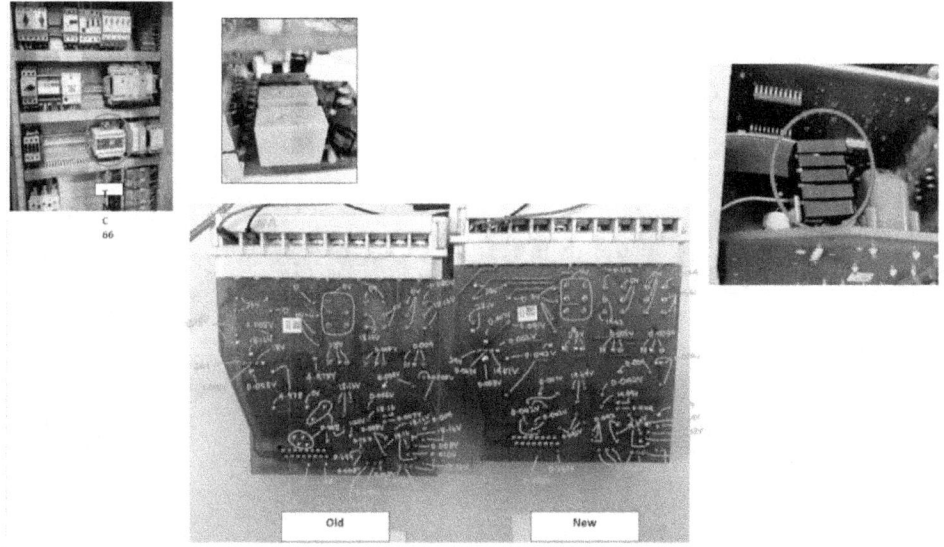

Figure 31: Voltage regulator failure from Chan Hui Ling & Foo Pei San Capstone project report.

This example relates to a control system shown top left in Figure 31 that de-activates a safety barrier at the correct time to allow pallets of containers to be fed into a depalletiser. At all other times the safety barrier stops the machine if activated to prevent operators from entering the machine.

The control unit had failed, and small burn marks are observed on some components shown top center.

Under power the voltage regulator on one PCB shown top right in Figure 31 was found to be arcing.

A detailed check of the voltage at each component comparing the PCB with a new one, as shown in the lower center of Figure 31, revealed that the voltage regulator was faulty, it's function being to step down from 24V to 5V.

We suspect the failure was caused by a voltage surge or fluctuation from elsewhere, but had to classify the failure cause as unknown, though not random.

7.4 TRANSMITTER CARD FAILURE

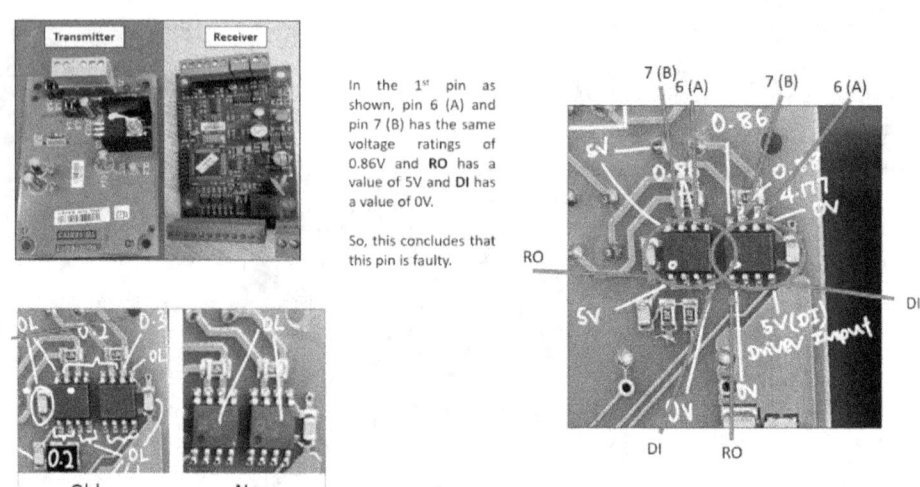

In the 1st pin as shown, pin 6 (A) and pin 7 (B) has the same voltage ratings of 0.86V and RO has a value of 5V and DI has a value of 0V.

So, this concludes that this pin is faulty.

Figure 32: Transmitter card failure investigation from Chan Hui Ling & Foo Pei San Capstone project report.

In the case of the transmitter card shown top left in Figure 32, the failure effect was a breakdown of the bottle filler.

Investigation found the failure mode to be incorrect voltage on Pin 6a of an integrated circuit on the transmitter card shown on the right of Figure 32, meaning that the IC logic is incorrect. The correct IC logic was confirmed by comparison with a new IC, shown bottom left.

The soldered connections to the IC appeared damaged and this may have occurred due to water from condensation or a failed repair attempt. Therefore, it was classified as an external cause.

7.5 BOTTLE SENSOR FAILURE

Figure 33: Failed bottle sensor from Chan Hui Ling & Foo Pei San Capstone project report.

The bottle sensor shown in Figure 33 failed to detect the presence of bottles. On inspection it was found that it had been packed with grease, probably in an attempt to prevent water from entering the sensor. However, the amount of grease was so great that it caused the internal wires to disconnect. We classified the failure as Human Error.

7.6 CPU FAILURE

Figure 34: Failed CPU unit from Chan Hui Ling & Foo Pei San Capstone project report.

The main CPU unit of a PLC controller is shown in the upper part of Figure 34. The unit had operated for 16 years before failure. Examination discovered thermal melt down of the IC as shown in the lower right of Figure 34. Further examination found that the cooling fan in the cabinet that holds the CPU card and other I/O cards is not operating. Therefore, the failure of the IC is external, caused by the failure of the cooling fan.

7.7 THE FINDINGS OF CHAN HUI LING & FOO PEI SAN

MECHANICAL COMPONENT FAILURES

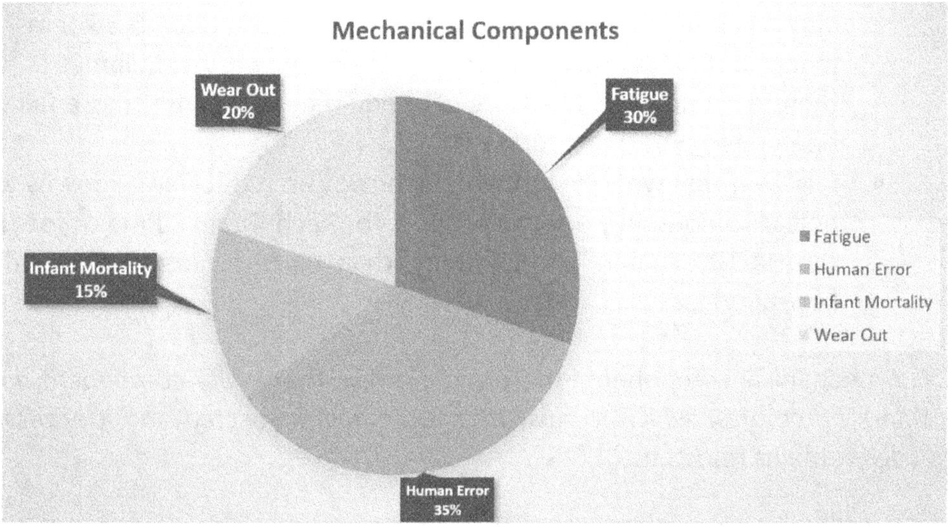

Figure 35: Failure causes from 20 mechanical failures at 2 breweries from Chan Hui Ling & Foo Pei San Capstone report.

Having examined the failure causes that occurred over 6 months at 2 breweries, we analysed the mechanical and electrical failures separately. We introduced the failure category of Human Error as there were several instances of incorrect actions on the part of Technicians or Operators, such as inserting grease into the bottle sensor as mentioned above. Another failure was due to the incorrect positioning of a sensor after a maintenance intervention.

It is noticed that wear-out failures tend to occur to flexible gaskets or seals which are the most frequent cause of this type of failure. For fatigue failures there were cases of loose screws or brackets most often caused by vibration (preventable with different fasteners).

There were no Random failures or Break-In failures. We did not assign a Bathtub failure pattern as this is a combination of the Wear-Out and the Infant Mortality patterns.

Our learnings from the analysis are:

- If you combine Infant Mortality (15%) and Human Error (35%) this accounts for 50% of failure causes, and this is not so far from the 68% for Infant Mortality found by Nowlan and Heap.
- For mechanical failures we find that 50% are related to operating hours and 50% are not. This is somewhat different from Nowlan and Heap finding only 11% of failures linked to operating hours. It must be taken into account that our sample size is much smaller as this was only a 6-month study. It could also be that the aircraft systems are much more carefully designed than brewery equipment, and therefore one can expect brewery equipment to suffer more failures related to operating hours.
- It is clear that with 35% Human Error we can reduce breakdowns with better competency and procedures for Technicians (a major focus of the HACS), Human Error is more than twice as frequent as Infant Mortality).

For mechanical components the basic premise that *every component has a certain lifetime after which it must be overhauled, inspected and if necessary replaced* remains rejected.

ELECTRICAL COMPONENT FAILURES

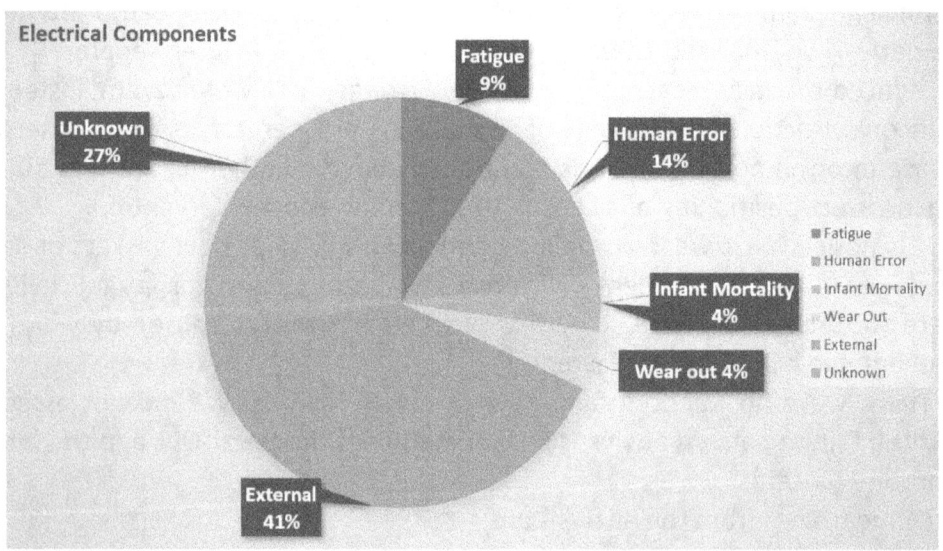

Figure 36: Failure causes from 22 electrical failures at 2 breweries from Chan Hui Ling & Foo Pei San Capstone report.

For the electrical components (Figure 36) we created the category of External failures, those that we know were caused by another component or external event. The examples given above of failures of the CPU and control panel due to inoperable cooling fans are illustrations of this, and also some failures due to water ingress.

It should be noted that the breweries are both located in Asia with high humidity and that temperatures in electrical cabinets can easily exceed 40˚C.

We also added a category "Unknown". There were several occasions where we could not identify the cause, so we record it as unknown instead of random, though we suspect many of these to be due to external factors. The category Unknown replaces the Random category, as we believe that every failure has an assignable cause and is never random. Some of the examples shown above would previously be categorized as random but the detailed investigation shows that there is an assignable cause to every failure.

Human error is a significant failure cause at 14% and Infant mortality is very low at 4% (It was 68% for Nowlan and Heap). One reason for the high infant mortality found by Nowlan and Heap could be that major overhauls/teardowns of the complex aircraft systems were the general practice at the time of their study.

Our learnings from this analysis are that many (at least 41%) of electrical failures can be prevented by taking care of the external conditions, especially cooling fans, water protection etc....

COMBINED RESULTS: MECHANICAL AND ELECTRICAL

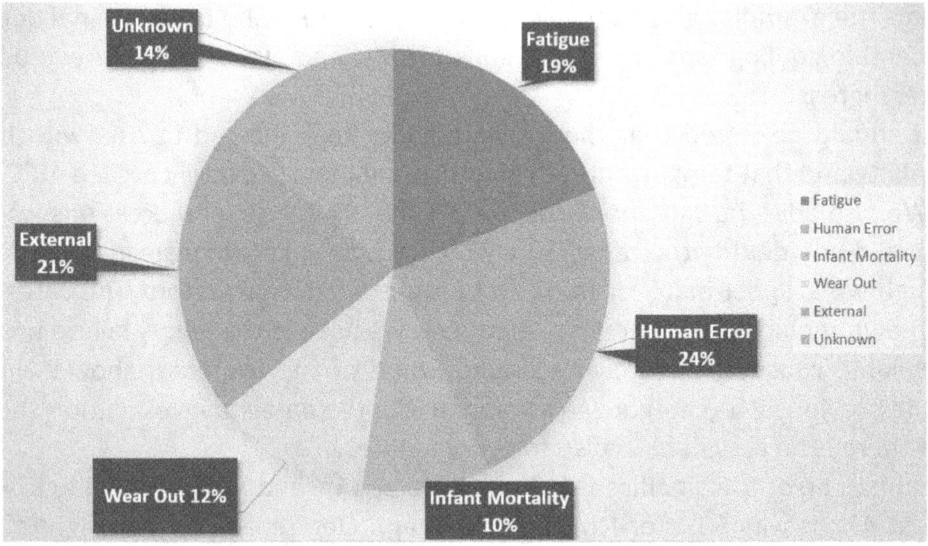

Figure 37: Failure causes from all 42 failures at 2 breweries from Chan Hui Ling & Foo Pei San Capstone report.

In Figure 37 we see the combined results for electrical and mechanical components. This is a useful perspective because Nowlan and Heap did not analyse electrical and mechanical components separately.

The research of Nowlan and Heap was more extensive than ours, our sample size is definitely much smaller as our study was only for 6 months.

However, we do get some insights that we can apply to the maintenance of a Brewery or beverage factory:

- Human error is a significant cause of failure. In our industry it accounts for 24% of failure causes.
- Analysis supports the hypothesis that there are no random failures. For many electrical failures we may not know the root cause, but detailed analysis can reveal a great deal about the failure mechanism that occurred. 14% of failures remain unknown, which is exactly the same percentage as Nowlan and Heap assigned to random failures.
- We did not have any failures that followed the break-in pattern.
- Fatigue and Wear-out occur at much higher rates than found by Nowlan and Heap. Overall, 31% of failures could benefit from time-based interventions. This may reflect that Engineering design is surely more meticulous in the aircraft industry.

- Infant mortality occurs at a much lower rate than found by Nowlan and Heap. This may be because we are not carrying out such frequent tear down overhauls as were common in the aircraft industry at the time of the Nowlan and Heap study.
- The External failure category is significant at 21%. It relates entirely to electronic components. At the time of the Nowlan and Heap study aircraft electronics were very rudimentary, using components such as relays, transistors and even valves/vacuum tubes that are more likely to follow an Infant Mortality failure pattern.

We did not analyze sufficient components to examine if there was a Bathtub curve separate to the wear out and infant mortality failures.

COMPARISON OF FINDINGS

Failure pattern	Nowlan & Heap	CHAN HUI LING & FOO PEI SAN
Bathtub	4%	
Wear-out	2%	12%
Fatigue	5%	19%
Break-in	7%	
Random/Unknown	14%	14%
Infant Mortality	68%	10%
Human Error		24%
External		21%

Clifford Jones

CHAPTER EIGHT: MAINTENANCE STRATEGIES

To guide you in thinking about Maintenance Strategies, I will summarize the relevant points of the previous chapters as follows:

FAILURE/BREAKDOWN DEFINITION

A functional failure is when a Maintenance Significant Item is unable to fulfill it's designed function(s) causing a breakdown or reduced performance of a production asset.

A breakdown is when a production asset is unable to produce products in specification due to one or more failures of Maintenance Significant Items.

FINDINGS OF CHAN HUI LING & FOO PEI SAN

- Many electrical failures can be prevented by taking care of the external conditions.
- Human Error is a significant cause of failure.
- Analysis supports the hypothesis that there are no random failures.
- We did not have any failures that followed the break-in pattern.
- Fatigue and Wear-out occur at much higher rates than found by Nowlan and Heap.
- Infant mortality occurs at a much lower rate than found by Nowlan and Heap.

8.1 MAINTENANCE STRATEGIES OF OTHER AUTHORS:

(Do note that I tend to use the terms maintenance strategy and maintenance action/activities interchangeably).

In this section all of our maintenance strategies are Preventive Maintenance tasks, except where we allow for Run to Failure which is then followed by a Corrective Maintenance task.

According to Smith, (Smith, 1993, p. 10), *"Preventive Maintenance is the performance of inspection and/or servicing tasks that have been preplanned (i.e.: scheduled) for accomplishment at specific points in time to retain the functional capabilities of operating equipment or systems"*.

"Corrective maintenance is the performance of unplanned (i.e.: unexpected) maintenance tasks to restore the functional capabilities of failed or malfunctioning equipment or systems".

The reasons to carry out Preventive Maintenance rather than Corrective Maintenance are:

- To prevent failures from occurring.
- To detect the onset of failure (P-F interval).
- To discover a hidden failure.

NOWLAN AND HEAP

Nowlan and Heap propose 3 maintenance actions or strategies:

- Scheduled INSPECTION of an item at regular intervals to find any potential failures.
- Scheduled REWORK of an item, at or before some specified age limit.
- Scheduled DISCARD of an item, at or before some specified age limit.

(They mention a 4th, scheduled Inspection of an item with a hidden function, but it is still a scheduled Inspection).

MOUBRAY

In his RCMII decision diagram, Moubray identifies the following tasks:

- Scheduled on condition task.
- Scheduled restoration task.
- Scheduled discard task.
- Scheduled failure finding task.
- Redesign.

SMITH AND MOBLEY

Smith and Mobley, in "Rules of thumb for Maintenance and Reliability Engineers" (Mobley, 2008, p. 43) have the following maintenance strategies:

- Lubrication.
- Inspection.
- Restoration.
- Discard.
- Redesign.

In the beverage industry we do not need to differentiate between restore and discard. Our maintenance program (HACS) focuses on the shop floor maintenance activities only and usually we exclude the activities that are most often executed off-line in the workshop or by a third party. So, if a motor fails it is replaced with another operational one. This unit comes from the spare parts store, and it may be new, or it may have been restored by an external specialist. This is explained further in the example given in Section 13.1.

At the shop floor, we only carry out a time-based replacement (discard task), without worrying where the replacement part came from or whether it was restored or new (It is quite understandable that one should care about this in other industries like aerospace).

So, for us, the restoration task is not required, it is always a replacement.

When we started building the HACS, we also took the decision not to carry out any redesign. I felt that if we started to redesign machines to make them less prone to failure or easier to maintain, we would never finish the project.

Whilst Smith and Mobley (Mobley, 2008, p. 43) added Lubrication to the tasks defined by Nowlan and Heap and Moubray, in the sources I refer to above there is no category for calibration (as mentioned earlier, Bloom devotes a chapter to RCM for instruments (Bloom, 2006, p. 181)). In our industry we use many process sensors that need frequent calibration, and this is an important maintenance task. For example, a temperature sensor needs to be checked whether it is giving the correct signal relative to the actual temperature. You might feel that this is a form of inspection, but I prefer to categorize it as calibration as it is carried out with very special tools and by a specialized Technician. Refer to Chapter 15 for more information on Calibration.

In the modern world we also have to include Condition Monitoring (CM). Condition Monitoring is the process of monitoring the components of an asset using vibration, temperature or other sensors in order to identify a pending failure (P-F interval). We will go into this along with Predictive Maintenance in Chapter 20.

And finally, we have the option not to do any maintenance. If the part is not really important, or very unlikely to fail and if so with a low consequence, you can elect to apply "Run To Failure". In this case there is no maintenance activity until the component fails. An example could be the radio in your car. It is nice to have, but not really important to the operation of the vehicle. However, it still appears in the maintenance schedule as we will be needing the part number when this unimportant item does fail:

This leads us to the following maintenance strategies, upon which our HACS system was built:

8.2 MAINTENANCE STRATEGIES USED IN THE HACS

- INSPECTION
- TIME BASED REPLACEMENT
- LUBRICATION
- CALIBRATION
- CONDITION MONITORING
- RUN TO FAILURE

The above tasks can be programmed into the Maximo CMMS system (except for Condition monitoring). We can define Maintenance Strategies as:

MAINTENANCE STRATEGIES are the targeted maintenance activities carried out to prevent the (functional) failure of a Maintenance Significant Item.

INSPECTION

An Inspection task is a periodic task repeated at a certain frequency, and should always contain a standard for the inspection, as well as a description of how to carry out the Inspection. It is referred to by Nowlan and Heap (Nowlan, 1978, p. 51) as an "on-condition" task.

For example, to measure the stretch on a chain, we use a certain type of chain-stretch gauge, and we have a standard that the stretch must be less than 3% (Figure 38).

In the Inspection task it is also necessary to state what should happen if the Inspection fails to meet the standard. In the case of a chain, it would be to replace the chain AND sprockets with new ones.

Figure 38: Inspecting chain stretch with a gauge, author's own collection.

I often see inspections written without a standard, such as "check the motor". It is essential to specify exactly what is to be checked. One of the pre-requisites set when I review any Planned Maintenance schedule/ Job Plan to be included in the HACS is whether every Inspection has a clear standard.

It is always preferrable to use a gauge rather than to make a specific measurement, as this reduces the risk of error. For the Inspection in Figure 38 it is possible to measure the stretch with a vernier versus a new chain and calculate the percentage stretch. This is time consuming and introduces a significant risk of error in either the calculation or the measurement. The chain gauge we use has windows on the device that shows the Technician directly the percentage stretch when he/she measures the specified number of links.

For the inspection task to be effective in identifying a pending failure, it has to occur in the P-F (Pending Failure) maintenance window (refer to Figure 26 in Chapter 6). Moubray defines the P-F interval as the interval between the occurrence of a potential failure and its decay into a functional failure in Chapter 7.2 of his book (Moubray, 1991, p. 145). The inspection must be carried out at intervals less than the P-F interval to have a possibility to detect the pending failure in time, and the general rule of thumb is for the Inspection interval to be not more than 50% of the P-F interval.

It is difficult to establish the P-F interval, as you need to know when the potential failure starts to become observable by skilled Inspection, and then the time period until the failure occurs. When the failure starts to be observable can be linear and predictable for a failure that is related to wear and tear but will often be unpredictable (or could be exponential) for a failure related to fatigue or corrosion.

The best source of information will most likely be the Technicians or Operators who look after the asset, they may be able to indicate the approximate time period of the P-F interval, and help to establish if it is weeks, months or years. Take care to distinguish between the time that the component has been in operation and the time from which the pending failure can be measured or observed. Spare parts replacement data can also be a good indication, as can data on previous failures.

Recently on a can seamer we experienced a failure of a knock-out rod, of which there are 12 in the seamer. The failure caused extensive damage inside the seamer that cost 30 000 USD to repair. The knock-out rod was replaced by the OEM in a major maintenance intervention some 4000 hours of operation earlier, and the next major maintenance was due in another 6000 hours. There was no interim inspection recommended by the supplier. Three more rods were found to be partly worn, the remaining 8 were OK.

What we know from this is that the P-F interval is not uniform for these parts but can be as low as 4000 hours. The supplier expects it to be over 10 000 hours and applies a Time-Based Replacement strategy. This is clearly the incorrect maintenance strategy, as the consequence of the failure is quite expensive, so we have added an Inspection at every 2000 hours.

Later in this chapter we examine in detail at how to choose the maintenance task, but at this stage do note that in general I have found inspections to frequently be the default choice of Engineers building a Planned Maintenance system (without proper structure). This then results in an excess of inspections, and in some operations, there is a very large team of technicians carrying out many different inspections every day, leading to a high maintenance cost.

I often ask my students, if they are seated in a plane waiting to depart and they look out of the window and see lots of Technicians with toolboxes under the aircraft checking and measuring things, how confident will they feel to stay on the plane?

The case of the seamer above is an exception, inspections should generally be eliminated as much as possible because they are expensive in manpower. They should only be used when the cost of replacement of the component is higher than the cost of the inspections of the component over its expected lifetime. Otherwise, it is cheaper to go for the next option:

TIME BASED REPLACEMENT

This involves the replacement of a component based on operating hours, operating cycles or calendar time passed, regardless of condition.

The most common mistake made with Time Based Replacement is to also Inspect the part at the same time, with the consideration to possibly delay the replacement. This should not be done as there may be a pending failure of the part that is not easily detected, such as the perishing or hardening of a seal.

Time Based Replacement is appropriate for components that follow the bathtub or wear-out failure patterns (11% according to Nowlan and Heap or 31% according to Chan Hui Ling & Foo Pei San, Chapter 7), and in this case we want to replace the part just before it's point of wear-out failure. Of course, identical components will not all wear-out at the same time, so there has to be some buffer in the time chosen to make the replacement.

In building the HACS, we ask the Technicians for the lifespan of the previous failed component, which sometimes we can get from the store's records. If we know that a bearing seems to last about 5 years before failure, then we can set the time-based replacement at 4 years. This is something that is easily adjusted based on feedback, to change the frequency of the time-based replacement is an easy update in the CMMS.

Time Based Replacement is an important strategy for low-cost items.

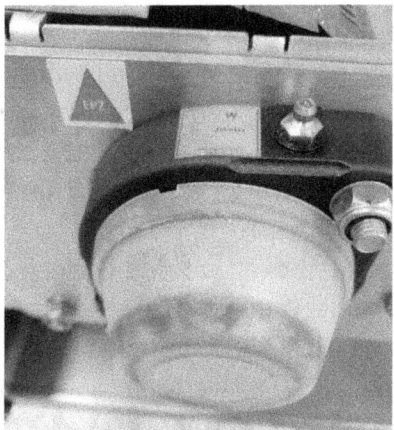

Figure 39: Conveyor bearing, from author's own collection.

Conveyor bearings are very numerous in a beverage factory, there might be 400 bearings like the one in Figure 39 on a single packaging line. Each one might cost about 20USD. To Inspect every bearing periodically would be a huge task, and over the lifetime of the part the inspection cost would be higher than the replacement cost. Actually, inspecting a bearing like this is not easy, depending on where it is installed it may not rotate fast enough for vibration analysis to be effective, and it would be ridiculous to try to dismantle it to Inspect the condition of the ball race.

So, for smaller bearings the correct strategy will be Time Based Replacement, and depending on the application, (Load, speed, environment) this is typically a 4- or 5-year replacement. So called "sealed for life" bearings last about 2-3 years in our applications.

For some electronic items like sensors or photocells we cannot predict when they might fail. However, we apply a Time-Based Replacement after ten years for most photocells and sensors, assuming that by then a more advanced version of the device can be installed. This can be considered optional.

LUBRICATION

The aim of Lubrication is to reduce friction between moving parts, because friction causes wear.

Lubrication tasks in the HACS include checking gearbox oil levels as well as the greasing of bearings, changing oil in gearboxes and inspecting it for indications of deterioration. When writing the Planned Maintenance schedules for each MSI, we always consider if it is a moving part and then what lubrication is needed. All moving parts need Lubrication.

Chapter 16 is dedicated entirely to Lubrication, and as described in section 16.1, we usually follow the manufacturer's recommendation with respect to Lubrication.

I have found that in many of our breweries "check the gearbox oil level" is a Planned Maintenance schedule issued every 3 months for every gearbox as a default. It seems to quite often be one of the first Planned Maintenance schedules that Engineers like to write if they are not approaching Planned Maintenance development in a structured way.

For smaller conveyor gearboxes we don't check the oil level so frequently anymore, based on the expectation that if the gearbox is leaking oil, it will be seen on the floor as an oil puddle, and someone will then take a corrective action (an operator should raise a Corrective Maintenance tag or job card request in the CMMS). Here there is an important interface between maintenance and operations, and this can also be included in a weekly CILT (Cleaning, Inspection, Lubrication and Tightening) check conducted by the operator.

CALIBRATION (See Chapter 15 for more details)

Calibration is the comparison of measurement values delivered by the device under test with those of a Calibration standard of known accuracy.

In the brewing process we need to accurately control the temperature at various points to get the natural enzyme reactions that we need, and in fermentation the correct temperature will determine the fermentation rate and flavour characteristics as well as the consumption of cooling energy.

In filtration we need to measure carbonation level, haze/clarity and alcohol content as part of automated processes that are critical to final product quality.

It is important to understand the difference between Calibration and Verification. Calibration compares measurement values delivered by the device with those of a calibration standard. Verification is an internal check of the instrument that ascertains if all functions are working correctly, and that there are no fault conditions, but it is not calibration.

The frequency of the Calibration depends on the impact of any instrument variation. If the impact is high, we might calibrate quarterly, if lower then every 6 months or one year.

CONDITION MONITORING:

Condition monitoring (CM) is the process of monitoring a particular condition in machinery (such as vibration, temperature, etc.) to identify changes that could indicate a developing fault, and like Inspection it needs to take place in the P-F part of the failure curve.

Oil analysis is one form of CM where we take an oil sample from a gearbox and a specialist company will analyse the sample for metal particle content in the oil. One can argue that oil analysis is a type of Inspection, as it is not a continuous monitoring, which is normally part of the definition of Condition Monitoring.

Our brewery in Cambodia is leading the implementation of continuous Condition Monitoring, working with IFM (a German automation company) we have installed Condition Monitoring sensors for temperature and vibration on our larger pumps, cooling compressors and CO_2 compressors.

Condition Monitoring does not remove the need to have an extensive Planned Maintenance system with a well-structured hierarchy, as you still have to know which components are to be monitored, where they sit in the machine structure, their function and the effect of their failure.

Hardwired sensors are linked to data modules which are connected by Ethernet to the monitoring system.

We only apply Condition Monitoring to the more expensive and critical assets. At the moment the general rule is that we apply it when an asset is not replaceable with an off-the-shelf item from the spare parts store.

For more information, please refer to Chapter 20.

RUN TO FAILURE

This strategy means we do not maintain the component in any way, we just replace it if/when it fails. By definition, if a component is on a Run To Fail strategy, we should have spare components in the store, and we will take corrective action immediately after the failure.

Can a component be an MSI and still have a Run to Fail strategy? Yes it can! If it is an MSI (see next section) then we believe it is an item to be included in the CMMS so that we know the part number and other information, but it can be that the failure is not critical and will not affect safety or operation, so we can allow RTF, or it can be that it is somehow important but we have no possible way to predict the failure.

In the HACS I have allowed very few run to failure strategies to be included, my personal benchmark is a maximum of 5% for any asset, but there is no scientific basis for this standard, it's just based on my own intuition.

Run To Failure is applicable either because we have no idea when the component will fail, but we can manage that failure when it occurs (perhaps with redundant systems), or because the component has a really minor function (like the car radio mentioned earlier), or because the failure is very unlikely to occur.

A hand operated ball valve on a water cleaning point is an example (Figure 40). They are infrequently used (a few times per day) and are extremely reliable, so this component will probably outlast the brewery before a failure, and if it does fail then the consequence is low, either a small leakage or we cannot clean in that particular place. However, every MSI failure has to be followed by a corrective repair action, the item must not be left in a failed state.

If it can be left in a failed state, then it is not needed and should be removed.

Figure 40: Hand operated ball valve, from author's own collection.

8.3 THE MAINTENANCE SIGNIFICANT ITEM (MSI)

Before we go on to describe how to choose the maintenance strategy, we need to understand the concept of the Maintenance Significant Item (MSI), and unfortunately it is not an easy concept.

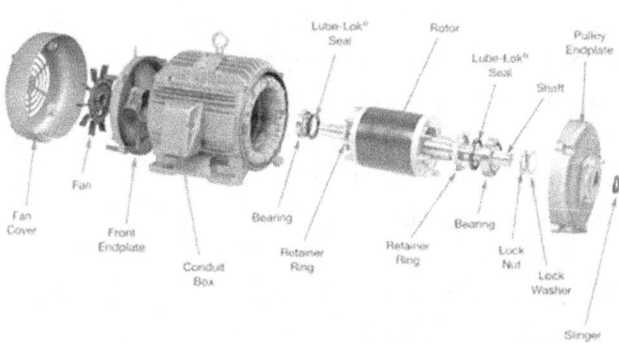

Figure 41: Maintenance Significant Item Challenge, from author's training materials.

In my training courses I show the picture in Figure 41, and I ask students which are the Maintenance Significant Items (MSI) on the small motor. There is a wide variance in answers, but generally they miss the correct answer, which is that in our case, in a brewery, the whole motor is the MSI.

This is because we have many hundreds of these small conveyor motors (sized about 0.75kW) and we don't repair them ourselves. Whether they have failed or they have reached a Time-Based Replacement point, we just swap the motor with a new one or one that has been overhauled, in the case of overhaul it is usually done by a third party.

But for a larger motor, we might well dismantle it and change the bearings on a periodic basis, or more likely we will apply Condition Monitoring.

The point is that what may be an MSI for one industry can be an assembly or sub-assembly elsewhere. It can also be affected by which of the parts in an assembly can be physically replaced. In the case of a small pneumatic piston or solenoid, the supplier will provide a complete replacement unit, rarely will they supply the internal components of such a device. Even more so for electronic parts, we almost never replace soldered parts on a printed circuit board (PCB), we just replace the whole board.

Smith defines a part (or piece part) as the lowest level to which equipment can be disassembled without damage or destruction to the item involved (Smith, 1993, p. 58), but as in the case of the motor an MSI can consist of several piece parts that can be disassembled.

Cost is also a factor. Recently I had to persuade our filler supplier, to change their spare parts availability. The can filler has a control assembly on each filling valve, and when they fail occasionally, almost always it is one of the 3 solenoid valves in the control assembly that fails. (The root cause is minute dust particles, so we are installing better air filters in the cabinet). We used to replace the solenoids, but the OEM solenoid supplier withdrew the solenoid valve availability and made it exclusive to the filler supplier, so we could not buy it anymore. A 60USD replacement of a solenoid valve from the OEM became a 600USD replacement of the whole control assembly from the machine supplier. Fortunately, after discussion both suppliers agreed to allow us to purchase just the solenoid again and the MSI moved from being the whole control assembly back to being the solenoid valve on the assembly.

In the development of the HACS, one of the first things we had to clarify is what is the MSI? How far will we drill down into the machine hierarchy to build our Planned Maintenance schedules and where do we stop?

As already mentioned in Section 4.1, I thoroughly researched the available texts and I could not find any guidance from either Nowlan & Heap (Nowlan, 1978), Moubray (Moubray, 1991) or Smith & Mobley (Mobley, 2008).

So, we had to develop our own rules, and these rules are absolutely essential to guide us in how many Planned Maintenance schedules we must develop and to what component depth.

Our starting point is that it can never be that an assembly or sub assembly that contains different replaceable components can be the MSI. The different components will have different failure patterns and therefore we have to write Planned Maintenance schedules at the component and not the assembly level.

We define how much we dismantle a particular item based on the spare parts that the supplier normally provides. My definition of the MSI is:

The lowest level component or sub assembly, that is likely to cause a breakdown of the asset in its useful lifetime, to which we apply a maintenance strategy.

It is either the part that is kept in the stores as a spare, or the lowest level of parts components available from the supplier.

I have seen instances where breweries and equipment suppliers take the Bill of materials for a machine, that is the entire list of every single component, and try to have a Planned Maintenance schedule for every single item. In the case of aerospace or industries with serious failure consequences, this may be necessary. As mentioned in Chapter 5, this was also the approach of the supportability project and led to lists of about 7000 items in the FMEA with many hundreds of MSIs incorrectly identified.

We don't have the time or resources to make a Planned Maintenance schedule for every nut and bolt in our industry, and we don't need to. Conveyor frames, guiderail supports, machine cabinets and body, access doors, perimeter fencing, access platforms are all items that are very unlikely to fail in the lifetime of the machine and so we do not consider them to be an MSI.

We have a lot of pipework in a brewery, but the pipe itself is not an MSI. The sensor on the pipe or the control valves will be MSIs, or the gasket at the joint, but never the pipe itself because it will itself almost never fail (Possibilities could be internal corrosion or blockage by scale etc. but in a brewery, we have strict control over the liquids in our process, so this is very unlikely). We do incur scaling of heat exchangers etc., but I consider that to be an external cause rather than a failure of the MSI.

In general, the rule of "it's the part you have in the store" works well for defining the MSI.

Building the Failure Mode and Effect Analysis at the MSI level is absolutely essential in order to correctly define the maintenance strategies that need to be applied to prevent each failure. If your FMEA contains phrases like "valve failed" or "pump fails to deliver volume", that cannot guide you to decide on the correct maintenance strategy unless you know the component that failed (Bearing. Spring, seal etc..).

8.4 DECIDING THE MAINTENANCE STRATEGY

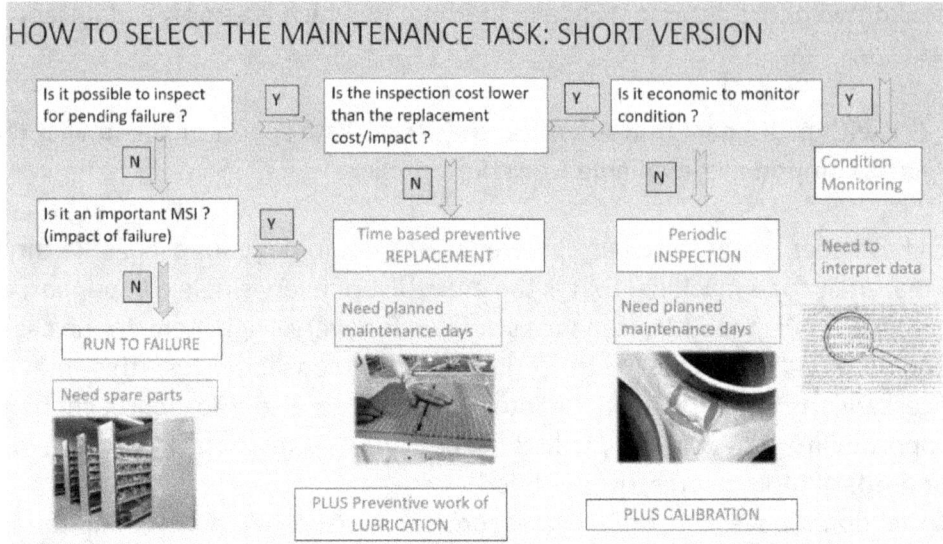

Figure 42: Deciding the maintenance strategy, from author's training materials.

I ask every Engineer that I train to print the slide in Figure 42 and keep it on their desk. Whether we are carrying out a Failure Mode and Effect Analysis, or a 5-Why, or just writing a schedule for a particular failure, once we have determined the failure cause (aka Failure mode, see next Chapter), we have to then choose the most appropriate maintenance strategy.

Moubray (Moubray, 1991, p. 198) Regan (Regan, 2012, p. 104) and others have much more complex decision diagrams. The above is designed to be simple and straightforward for a Brewery Engineer to be able to follow.

We start in the top left with:

IS IT POSSIBLE TO INSPECT FOR PENDING FAILURE?

This first question is the most important in deciding the maintenance strategy: Is it possible to inspect for pending failure? In other words, is there a P-F interval in the failure curve (Figure 26, chapter 6) where we can use instruments to detect the pending failure? Can the pending failure be detected by measuring wear, size, thickness or some other physical feature?

If we have an electrical component, most likely the answer is no. For many mechanical parts the answer should be yes, as we can measure wear in a cylinder, stretch on a chain or wear on a sprocket or bearing, or inspect for cracks in the case of fatigue. At this stage we do not concern ourselves how we will do that inspection, just whether it is practically possible or not.

If it is not possible to inspect for pending failure, we then have to decide if the MSI is important, is there a significant consequence to its failure? If there is, then we have to have a Time-Based Replacement, and if not, we allow Run To Failure. What is a significant consequence? It would be any safety issue, any environmental issue, any product process variation and any asset downtime that would be considered a significant consequence.

Bloom (Bloom, 2006, p. 53) designates Run to Failure if there is no safety, operational, regulatory or economic consequence as the result of a single failure, and the occurrence of the failure is immediately evident to operations personnel. The occurrence of the failure has to be followed with a Corrective Maintenance action and not left as-is.

If it is possible to inspect for pending failure, we next consider the cost of that inspection with the question:

IS THE INSPECTION COST LOWER THAN THE REPLACEMENT COST?

If the inspection cost is low relevant to the replacement cost, we will move to the right of the chart and consider Condition Monitoring as a final alternative before possibly opting for Inspection.

But if the inspection cost is higher than the replacement cost, we will move down to Time Based Replacement.

In our example of the seamer knock-out rod failure (earlier in this Chapter) the failure caused 30 000 USD of internal damage to the machine and a very strong likelihood of a long wait for replacement parts and consequent loss of production of several weeks. Inspection of the knock-out rods can be done fairly easily, and it might take two hours for all 12 to be accessed and checked. Two hours of Technician time every 2000 hours is going to cost less than the likely cost of another failure, so Time Based Replacement, as was specified by the supplier, is the incorrect option in this case.

Do note that we have to always remember to include the full number of inspections in the expected life of the component. Let's consider a common type of valve that we have in breweries, known as a double seat valve. Often these valves are installed in a valve matrix as shown in Figure 43, so we have to safely isolate the valve matrix and then there is a lot of disassembly required to remove them (and they are heavy). Then the individual valve itself has to be dismantled to get to the seals, one of the most important MSIs in the valve.

Figure 43: Double seat valve and valve matrix, from author's collection.

As the valve seal replacement kit will cost less than 50 USD, then the labour cost of carrying out the disassembly perhaps 3 times a year over about 4 years (as would be needed for an Inspection strategy) is going to be far higher than just replacing the seals every 4 years (which is what we normally do), so Time Based Replacement is the correct choice here and not Inspection. (The consequence of failure is hard to estimate. Usually, it is low because the valve is designed in a way to prevent the possibility of product contamination, a feature of the double seat design.)

The choice between Inspection and Time-Based Replacement is a balance between the consequence of the failure, the cost of the MSI and the cost of the Inspection.

IS IT ECONOMIC TO MONITOR CONDITION?

Given that the inspection cost is lower than the replacement cost, we can now ask the final question: "is it economic to monitor condition?".

To install a robust Condition Monitoring system for 6 or 10 fairly large assets like pumps and compressors will currently cost about 50k USD or more. This assumes 4 or 6 sensors on each asset, and that all the sensors are hardwired into a data recording and analytical system. (It is possible to install cheaper Condition Monitoring systems: many companies offer Wi-Fi based sensors for example, but in most cases, you will no longer own your data, so you need to evaluate the likely future cost of the analysis of your information and the access to it). Please refer to section 20.2 for more detail on implementing Condition Monitoring.

Currently we only go to Condition Monitoring if it is an asset that is not off-the-shelf, such as pumps of over 10kW, or compressors or generators that are large and complex. In the case of double seat valves one supplier has developed a version with built in sensors to detect leakage, it is quite expensive, but it could be of interest in the future.

As costs fall and networks improve, we will of course expand the number of devices using Condition Monitoring, and merge to a more integrated IoT.

At the moment, for most items when we have got to this point on the decision diagram, we will then decide to carry out a periodic Inspection.

8.5 ONE MSI WITH MULTIPLE FAILURE MODES

Valve seals are an example of an MSI that can have more than one failure mode depending on the environment or operating context.

We examined many valve failures in our Cambodia brewery and found two different valve seal failure modes, one being mechanical wear on the seal, the other being perishing of the seal. This is shown in Figure 44 with mechanical wear on the left-hand side and perishing shown on the right-hand side.

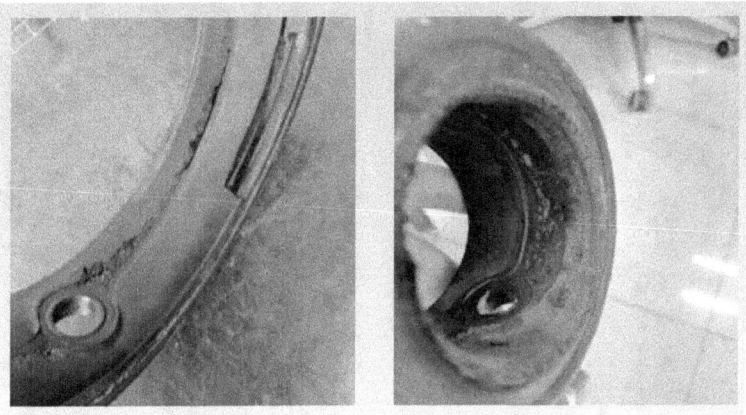

Figure 44: Different failure modes in valve seals, from author's own collection.

The mechanical wear is related to the number of cycles and will be the failure mode where the media inside the valve is not aggressive in deteriorating the flexible seal. However, where the media is aggressive (temperature or chemical concentration) then the seal may fail before the mechanical failure due to the perishing effect of the temperature or chemicals.

An automated valve in a mild environment may fail from the mechanical action on the seal after about 90 000 to 100 000 operations, but in a harsh

Clifford Jones

chemical environment of hot caustic it will fail in about 4 years from perishing of the seals.

So, there are 2 different failure modes requiring 2 slightly different maintenance strategies depending on the application of the valves. For mild conditions replacement is based on the number of operational cycles, for harsh conditions replacement is based on calendar time.

We follow a similar approach for pneumatic cylinders, where the OEM (Festo) specifies the design life of the cylinder in terms of the total stroke distance it can travel. Dividing the cylinder length into this number gives a number of operations of the pneumatic cylinder. We know how frequently the cylinder will be operated, so we can estimate when it will wear out. However, if the cylinder is used less frequently, then the seals my perish with time before they are mechanically worn out and we still need a Time-Based Replacement. For cylinders we calculate when it will wear out from operating cycles and use that if it is less than 10 years, if not we anyway replace after 10 years to avoid perishing of the seals.

8.6 SUMMARY OF THE MAINTENANCE TASK SELECTION CHART

I deliberately designed the chart to evaluate Time Based Replacement before Inspection, which is the reverse of the approach of Moubray (Moubray, 1991, p. 198) and of Regan (Regan, 2012, p. 104). My thinking was to make the Inspection the last item to choose, after all others have been rejected, to avoid the tendency I see in our Engineers to implement too many inspections as a default maintenance activity.

As mentioned on the chart, we need to ensure Lubrication and Calibration activities are included where needed, and we need to have Planned Maintenance days to carry out the Inspections or Time-Based Replacements.

So far, this chart has served us well, it is the foundation of the HACS and is used to decide the maintenance strategy for each MSI.

Moubray includes Default Maintenance Actions in the event that a proactive maintenance task cannot be found (Moubray, 1991, p. 170). These actions are to carry out a failure finding task in the case of a hidden failure, to opt for Run To Failure where there is a low consequence of failure (we already have this), or finally to redesign the equipment if the consequence of failure is serious and a proactive task cannot be found.

We decided in the beginning of the HACS development not to consider redesign, but this could (and maybe should) be explored in the future if we get to the stage of fine tuning the HACS system. Clearly these default actions would be necessary and important in high-risk industries such as aerospace or nuclear industries, but the use of the term "default" is confusing, in my view.

Given that we now know something about the types of failures, and the available maintenance strategies, we next need to have a process to evaluate how each component fails. This is called Failure Mode and Effect Analysis.

CHAPTER NINE: PRACTICAL FAILURE MODE AND EFFECT ANALYSIS

Failure Mode and Effect Analysis is a structured technique that describes the function and possible functional failure of an asset, system or process, identifies the failure mode that is likely to cause that functional failure and the failure effects associated with that failure mode.

It is a structured way of evaluating all the different ways that something might fail.

It should be remembered that it serves only one purpose, to help us identify the correct maintenance strategy to prevent failures and breakdowns, or as some authors prefer to put it, to prevent loss of function.

Failure Mode and Effect Analysis was published as early as 1949 in the US Department of Defense document "Procedures for Performing a Failure Mode Effects and Criticality Analysis" which has been updated several times since then, (it did not include Criticality in the earlier versions,) (US Department of Defence, 1980) https://elsmar.com/pdf_files/Military%20Standards/mil-std-1629.pdf

When Criticality is assessed, we have FMECA, Failure Mode Effect and Criticality Analysis.

Do note that it is not the only way of evaluating how things fail. 5-Why root cause failure analysis also plays an important part in this respect and is described in the next chapter.

Other techniques can be used, such as Fault Tree Analysis (FTA), an approach to examine an undesired state in a system that is used in high hazard industries such as aerospace, nuclear power, chemical and process industries (Fault tree analysis, 2020):

https://en.wikipedia.org/wiki/Fault_tree_analysis

Fault tree Analysis starts with an undesired outcome and then maps the relationship between faults, subsystems and design elements to create a logic diagram of the overall system and then calculates the probability that the undesired event will occur. It was developed by Bell Laboratories in 1962 to evaluate the Minuteman ICBM launch control system. It is used widely in civil aviation, is described in a FAA handbook, it is used by NASA and in the US nuclear power industry.

Fault Tree Analysis has no application in the Brewing and Beverage industry so it will not be discussed further.

9.1 OPERATING CONTEXT

Regan (Regan, 2012, p. 43) recommends that we define an Operating Context for the asset before commencing FMEA. This is a document of relevant technical information such as the scope of the FMEA, theory of operation of the asset, description of the asset, relevant operating conditions etc.

This is good advice and as a minimum we need to have available the manufacturer's documentation such as P&ID, layout drawings, maintenance manual and spare parts catalogue. The spare parts catalogue is essential to identify all the components that may incur a failure, those with "exploded" parts diagrams are the most useful.

Further, Regan considers the effect of the operating context on the failure mode. For example, if a part fails due to normal equipment operation, or if it fails due to extreme conditions. This has quite some relevance in the aircraft industry and Regan gives an example of 3 air tanker crashes that occurred due to structural failures when the aircraft were used in highly stressful fire-fighting roles instead of their designed cargo or tankering roles (Regan, 2012, p. 17).

The operating conditions and their effect are not so much of an issue in the Brewing and Beverage industry, as we have fixed facilities inside protective buildings, so environmental conditions are fairly constant, and in general the equipment is operated at a nominal rated capacity.

9.2 FMEA DEFINITIONS

The definition of the terminology used in Failure Mode and Effect Analysis is often problematic, even to Engineers, so I will attempt to explain it clearly.

Moubray in his RCM2 states:

A function statement should "*consist of a verb, an object and a desired standard of performance*". Such as "to pump water from the tank at not less than 800l/min".

Moubray (Moubray, 1991, p. 22) describes Primary Functions and Secondary functions, and describes functional failures as total, partial and hidden failures.

He states that "*a functional failure is defined as the inability of any asset to fulfill a function to a standard of performance which is acceptable to the user*".

Moubray gives several examples of functions and functional failures in his book, but they are all examples of complete assets or systems. His methodology and terminology is biased towards carrying out FMEA at an asset or assembly level. This leads to a different emphasis in the functional description (many sub functions then have to be described in the one higher level function) and makes the analysis complex. I found much more benefit in having a hierarchical structure well defined first, and then working at the component level to define the function of the relevant MSIs.

If we follow Moubray's approach, we get a functional statement for the asset such as the one above, to pump water from the tank at not less than 800l/min. Smith advocates to work at the entire system level (Smith, 1993, p. 55), but Bloom tries to work at the "component level" (Bloom, 2006, p. 112), though he does not drill down to the MSI, he considers assemblies to be components, and he structures his analysis by consequence of failure.

Clifford Jones

When we work at the MSI level, we have individual component functional statements, such as "to allow liquid flow in the pipe in only one direction" for a non-return valve on the same pump, and I find this to be much more relevant in developing a Planned Maintenance system. With an MSI level functional statement we can then develop the necessary maintenance strategy and Planned Maintenance schedule for the specific item, in this case the non-return valve.

I already proposed my own definition of a functional failure in Chapter 6:

A functional failure is when a Maintenance Significant Item is unable to fulfill it's designed function(s) causing the breakdown or reduced performance of a production asset.

Following on from this, taking into account the need to work in a structured hierarchy at a component level, here is my definition of the failure mode:

The failure mode is a description of the physical changes that caused the functional failure of a MAINTENANCE SIGNIFICANT ITEM.

Examples:
- Seized bearing.
- Loose impeller.
- Worn guiderail.
- Leaking seal.
- Jammed valve.
- Burned IC.
- Broken shaft.

In describing the failure mode, we focus on the failed MSI and describe the physical change that constitutes the failure of that MSI.

Continuing at the component level, my definition of the failure effect is:

The failure effect describes the impact that the MAINTENANCE SIGNIFICANT ITEM failure (failure mode) has on the assembly, system or asset of which it is a part.

9.3 UPDATED DEFINITIONS IN FAILURE ANALYSIS

Here are my updated definitions of a failure, an MSI and a failure mode and effect that we will use going forwards:

A FUNCTION is the process, action or task that the asset, system or MAINTENANCE SIGNIFICANT ITEM is designed to perform.

A FUNCTIONAL FAILURE is when a MAINTENANCE SIGNIFICANT ITEM is unable to fulfill it's designed FUNCTION(s) causing the BREAKDOWN or reduced performance of a production asset.

A BREAKDOWN is when a production asset is unable to produce products in specification due to one or more failures of MAINTENANCE SIGNIFICANT ITEMS.

A MAINTENANCE SIGNIFICANT ITEM is the lowest level component or sub assembly, that is likely to cause a BREAKDOWN of the asset in its useful lifetime, to which we apply a MAINTENANCE STRATEGY. It is either the part that is kept in the stores as a spare, or the lowest level of parts components available from the supplier.

MAINTENANCE STRATEGIES are the targeted maintenance activities carried out to prevent failure of a MAINTENANCE SIGNIFICANT ITEM. They are Inspection, Time Based Replacement, Lubrication, Calibration, Condition Monitoring and Run to Failure.

A FAILURE MODE is a description of the physical changes that caused the functional failure of a MAINTENANCE SIGNIFICANT ITEM.

A FAILURE EFFECT describes the impact that the MAINTENANCE SIGNIFICANT ITEM failure (failure mode) has on the assembly, system or asset of which it is a part.

9.4 HOW TO CARRY OUT FAILURE MODE AND EFFECT ANALYSIS

In chapter 5 I showed how complex FMEA can be, giving the example from the supportability project, which used 60 columns in total to carry out the FMEA.

At the core of FMEA we just need to describe the function, the functional failures, failure modes and failure effects.

In my training courses I use an electric kettle as an example, the type that switches off automatically, because this is an item that nearly everybody is familiar with.

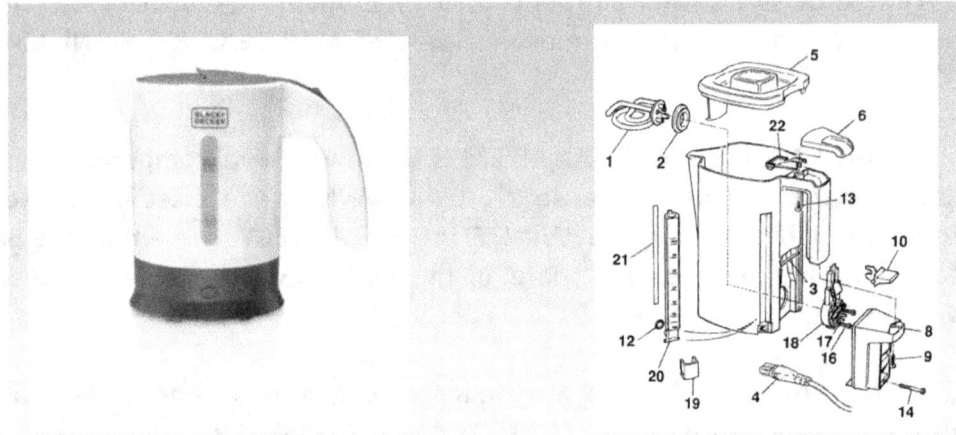

Figure 45, automatic kettle from author's training materials.

Having the exploded parts diagram as shown in Figure 45 helps us to consider the components that might fail and how they can fail. This visualization is important and is normally included in the HACS hierarchical structure.

Being able to see the components also tells us a lot about how the kettle operates.

Now we will carry out the FMEA for the whole kettle, noting that this is possible without a hierarchy for a simple device like this, but for a complex machine we would first need to breakdown the asset to assemblies and sub-assemblies before carrying out the FMEA.

For complex machines, we must always complete the hierarchical structure breakdown FIRST.

FUNCTION	FUNCTIONAL FAILURE (LOSS OF FUNCTION)	FAILURE MODE (CAUSE OF FUNCTIONAL FAILURE)	FAILURE EFFECT
	PRESENT	PAST	FUTURE
To raise the temperature of water to boiling	Does not heat water to boiling	Heating element burned out	Cold coffee
		Fuse blown	Cold coffee
		Thermostat not connecting	Cold coffee
		Cable burned	Cold coffee
	Partially heats water	Thermostat early disconnection	Luke warm coffee
To switch off automatically	Does not switch off	Thermostat fails to disconnect	Boils dry, fire risk
To contain the water for boiling	Water leaks out	Cracked kettle	Boils dry or shorts out.
		Perished seal	Boils dry or shorts out.

Figure 46: Failure Mode and Effect Analysis of a kettle, from the author's own training materials.

When we consider the kettle, the primary function is fairly obvious, to raise the temperature of water to boiling point. This is in line with our definition that a function is the process, action or task that the asset, system or Maintenance Significant Item is designed to perform.

But there is an important secondary function, which is to switch off automatically when boiling has been achieved.

You might also define another secondary function, to contain the water without leaking.

There could even be another function, to prevent the entrance of insects to the water, if you look in your kettle you should find a small sieve at the spout that prevents insects from crawling into the kettle. In failure analysis we always have to decide when to stop considering unlikely or unimportant failures, and the point that you do that will depend on the industry that you are working in and the consequence of failure.

Returning to the first function, to raise the temperature to boiling, we can then consider how it can fail to meet that function. It could fail completely, meaning that it does not work at all, or it could fail partially, heating the water to a lower temperature but not boiling. In imagining this partial failure, we have to know that there is a thermostat controlling the heating element of the kettle and it is therefore possible for this to occur. You can see the thermostat as component 18 in Figure 45, and here is where we might discuss in my training courses how a bi-metallic strip operates:

https://en.wikipedia.org/wiki/Bimetallic_strip (Bimetallic strip, 2012) (note that electronic thermostats are becoming more common and allow continuous heating on and off to a set temperature).

We have two potential functional failures for the first/primary function, to not heat the water to boiling at all, or to partially heat the water.

For the second function, to switch off automatically we can then have the functional failure of not switching off. We could also consider here a functional failure of switching off too early, but that is the same as the functional failure defined above of partially heating the water, so it is not needed.

And for the function to contain the water, then the failure would be for the water to leak out.

Once functions are defined, defining the functional failures is quite simple, it is usually just the opposite of the function occurring. Bloom advocates (Bloom, 2006, p. 99) that the specification of the functional failure is not needed as the functional failure is the exact opposite of the function. This is not the case in a partial functional failure, so in my view the functional failure definition is still needed.

Now we come to describe the possible failure modes, remembering that a failure mode is a description of the physical changes that caused the functional failure of a Maintenance Significant Item.

It is at this point that we must move to the component level (MSI), and consider how each MSI can fail, and here the parts diagram is essential.

Working on the functional failure "does not heat water to boiling", we have to look at the components and consider if they can fail in a way that causes this functional failure.

We should know from experience that a failure of the heating element (component 1) is a likely failure mode. So, this component is an MSI. A heating element is just an electrical resistance wire and the physical change that causes this failure is usually a complete breakdown of the resistance of the wire so that it fails to heat. Likewise, a blown/burned out fuse would cause a failure to heat the water. This is a physical change in the fuse whereby it has burned out and no longer conducts electricity. (Actually, for the fuse this is its primary function, it is designed to burn out in the case of an overload) The fuse is an MSI.

The thermostat (component 18) could cause this failure, if it remains in the open (disconnected) position and does not close to provide current to the heating coil. And the cable (Component 4) could be burned, or even not plugged in, which would cause the functional failure. The thermostat and the cable are MSIs.

I think that all possible failure modes are covered for this functional failure, but it could be that there are others.

The next functional failure is to partially heat the water, and this would only occur from an early disconnection of the thermostat (Component 18) (except in the case of a complete failure during the heating process, but then there is no heating at all on the next cycle). The physical change here in the component would be that the bi-metallic strip is not operating, perhaps the riveting of the strip is disconnected (fatigue failure), or the bending amount of the strips has changed due to age, or the internal contacts are corroded from arcing. But we do not need so much detail (The whole bi-metallic strip is an MSI, and we cannot usually dismantle it), and it is sufficient to say that it disconnects early.

The next functional failure, failure to switch off automatically, would again be due to the bi-metallic strip, in this case it stays closed and does not open to break the circuit. This is the failure with the most serious consequence, there are numerous cases reported of house fires caused by an automatic kettle not switching off, it typically boils dry, overheats, melts and catches fire.

Finally, we have the functional failure of water leaking from the kettle. Other than it being cracked, there is a seal on the heating element (component 2) that could fail (perish)and cause leaking, and also a seal (component 12) on the level indicator that could fail (perish) and cause leaking. These are also MSIs.

So, we now have a list of about 8 failure modes for the kettle:

- Heating element burned out.
- Fuse blown.
- Thermostat not connecting.
- Cable burned.
- Thermostat early disconnection.
- Thermostat fails to disconnect.
- Cracked kettle body.
- Perished seals.

Finally, we look at the effect of each failure. As you can see, mostly the effect is that I don't get my hot coffee (availability), but sometimes the effect can be quite serious, it boils dry, overheats and melts and catches fire. It could also be that there is an electrical short from water leaking.

Regan includes a flowchart in chapter 6.4 of her book (Regan, 2012, p. 79) to decide which failure modes to include in the analysis. According to her, the failure mode is included if:

- It has happened before.
- It is a real possibility.
- It is unlikely but has a severe consequence.
- There is an existing pro-active maintenance task.

In the case of Failure Mode Effect and Criticality Analysis, we would next consider the consequence of the failure mode on Availability, Safety, the Environment, Cost and Quality. Usually there is an arbitary score given to each of these factors, and then they are multiplied to obtain an overall criticality score.

This approach to criticality assessment is a fundamentally flawed process because you are multiplying scores that come from unrelated scales. There is no logic to multiplying a score for Safety consequence on an arbitrary scale by a score for Environmental consequence, also on an arbitrary scale, and using the product to decide criticality, but it is a commonly applied way of working in criticality analysis.

Bloom (Bloom, 2006, p. 74) advocates to classify failure modes as Critical, Potentially critical, Commitment (regulatory), Economic and Run To Failure. He follows practices developed in the nuclear power industry and uses the failure mode criticality to decide on the maintenance strategy, which is not logical.

It makes more sense to me that the maintenance strategy is decided by the physical characteristic of the failure mode (The physical changes that cause the failure), as that will determine whether an Inspection, Time Based Replacement or Condition Monitoring is even possible. If we want to consider criticality, that should then be built into the **frequency** of the chosen maintenance strategy.

9.5 DETERMINING THE MAINTENANCE STRATEGY FROM THE FMEA

The reason for carrying out the FMEA was to help use the MSI failure modes to determine the maintenance strategy (and then apply maintenance activities to reduce failures).

In order to do this we take each MSI, and it's expected failure mode, through the maintenance task decision flowchart that I introduced in Chapter 8.

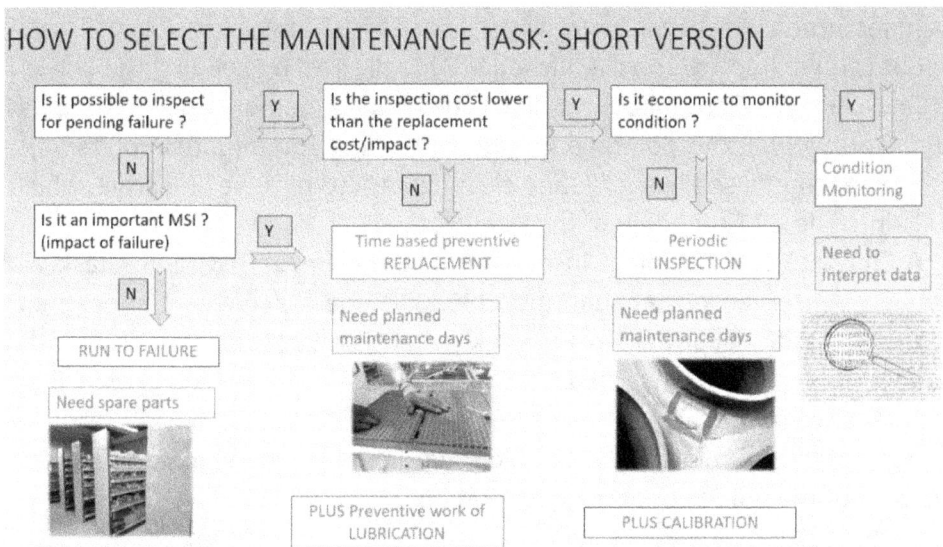

Figure 47: Maintenance Task Decision Flowchart from the author's training materials.

As it is a lengthy worked example, I have used capitals in the following explanation to refer to the questions in the Maintenance Task Decision Flowchart (Figure 47). We will review each Failure Mode of the kettle in turn:

FAILURE MODE: HEATING ELEMENT BURNED OUT

IS IT POSSIBLE TO INSPECT FOR PENDING FAILURE? It is possible to inspect after a failure has occurred, a burned-out heating element will have an infinite resistance. But it is not really possible to test for a pending failure, there is not usually any gradual increase in resistance or measurable deterioration. You could thermographically inspect the element for hotspots, but you are unlikely to find them in advance of a failure, as probably the P-F interval from pending failure to failure on the failure curve is very small (Refer to Chapter 6, The Failure Process). So, we choose NO.

Thus following the chart downwards we have to consider next IS IT AN IMPORTANT MSI? In terms of the function of the kettle it is important, so we choose YES, move to the right and we see that we need a Time-Based Replacement.

Now we need to estimate the time-period for the replacement. We can ask friends or family how long their kettles have lasted, and we might find that the lifetime is about 5 years, depending on how often it is used. In the brewery we would try to find out the average or expected life of the component from spares replacement records, sometimes we might have information from the manufacturer.

But can we improve on the lifespan? Actually yes, by keeping the kettle free of limescale we will reduce the load on the heating element and increase its life. (In the brewery at this point we should be thinking about lubrication tasks).

So, we might also have a routine cleaning task to remove limescale, which we build into the CILT schedule of the operator.

A CILT task is a Cleaning, Inspection, Lubrication or Tightening task with a frequency interval shorter than monthly. CILT checks/activities are carried out by the operator, usually against a weekly checklist. Every task that is monthly or less frequent than that should be programmed into the CMMS as a Planned Maintenance schedule.

As this task is 6 monthly, we would program it into the Planned Maintenance system but mark it for operator completion.

FAILURE MODE: FUSE BLOWN

IS IT POSSIBLE TO INSPECT FOR PENDING FAILURE? .....No, it is not, because a fuse will burn out due to unforeseen outside causes such as a power surge, or due to a short circuit in the kettle itself.

Following the chart downwards we have to consider next IS IT AN IMPORTANT MSI? In terms of the function of the kettle it is important, but we also have to consider that the potential "failure" is only from outside causes, so there is no benefit from a Time-Based replacement, so we can leave the fuse at Run To Failure and keep a spare fuse in the drawer. (In some respects, a blown fuse is not a failure, it has done its job to prevent damage to the equipment).

FAILURE MODES: THERMOSTAT NOT CONNECTING, THERMOSTAT EARLY DISCONNECTION, THERMOSTAT FAILS TO DISCONNECT

All 3 of these failure modes are very similar and refer to the same component, so I decided to group them together. We can group them together when there are not multiple causes of the same failure mode (For example, a seal my perish over time, or it may perish due to chemical action. In this case we might have 2 different maintenance strategies on one MSI for the same failure mode, this is explained in Chapter 8, but it is not the case here.) IS IT POSSIBLE TO INSPECT FOR PENDING FAILURE? .....no, it is not, unless you are going to dismantle the thermostat, which would probably destroy it.

Following the chart, we have to consider next IS IT AN IMPORTANT MSI?. In terms of the function of the kettle it is important, so we need a Time-Based Replacement.

As for many electronic parts, there is little that we can even try to do by way of inspection, though we might have some CILT checks to ensure that it is clean and dry.

As mentioned above, we probably will not know the real cause of failure of this part, but as explained in Chapter 7 there will be an assignable cause, it is not random.

Again, we have to estimate the lifetime of the part. As our friends tell us that a kettle lasts about 5 years, and this is a critical part that has several failure modes then we will apply the same 5-year Time Based Replacement.

FAILURE MODE: CABLE BURNED

IS IT POSSIBLE TO INSPECT FOR PENDING FAILURE? .....Yes, you can check the condition of the cable for cracks or pinching that may lead to it burning out. The cable shorting out internally would usually be caused by some sort of fatigue process, such as a kink in the cable.

Now we move to the right in the flowchart and ask IS THE INSPECTION COST LOWER THAN THE REPLACEMENT COST? Well, assuming that the power cable costs about 2 dollars, and if we need to pay a technician to carry out the inspection, maybe it takes 5 minutes and has to be done 4 times per year (We have to estimate the length of the P-F interval before the failure occurs), then the inspection cost is higher than the replacement cost, so we should have a Time Based Replacement, let's say every 3 years we will change the cable for a new one.

FAILURE MODE: CRACKED KETTLE

IS IT POSSIBLE TO INSPECT FOR PENDING FAILURE? .....It is possible that you can see cracks in the body of the kettle before they become large enough to cause a leakage. So, the answer is YES, and we move to the right of the chart.

IS THE INSPECTION COST LOWER THAN THE REPLACEMENT COST? Let's say in this case it is lower, mostly for the sake of argument and to demonstrate the use of the decision chart.

So now we move to the right again and ask IS IT ECONOMIC TO MONITOR CONDITION? In theory we can fit our kettle with sensors to somehow monitor each of the MSIs, but in practice, how would you check for cracks in the body?

You can have a level sensor but then it has to know not to give a false signal when you empty the kettle by hand. To check for failure of the thermostat would be even more complex, you would need several sensors, and some built in processor to ensure it has switched off at the correct temperature. In this case the Condition Monitoring would be more complex than the original device. This example demonstrates one of the challenges in Condition Monitoring and Predictive Maintenance, the sheer complexity of the monitoring and analysis needed. So, it is not economic to have Condition Monitoring on a 20USD kettle, and we answer NO and move down the chart to a periodic Inspection. We could assume that any cracks take a long time to develop to become complete leaks, so we could have an annual inspection here.

FAILURE MODE: PERISHED SEALS

IS IT POSSIBLE TO INSPECT FOR PENDING FAILURE? …..Generally, it is possible to inspect a seal for signs of perishing, but it is not a very reliable inspection. Perhaps there is hardening that you cannot detect, or small cracks that are not yet easily visible.

But we can be optimistic about the possibility of inspection, and move to the right in the flowchart and ask IS THE INSPECTION COST LOWER THAN THE REPLACEMENT COST/IMPACT? Seals are usually very cheap, so we can say NO, the inspection cost is not lower than the replacement cost/impact, and we will move down and have a Time-Based Replacement.

But, we already have a Time-Based replacement for the heating coil, and it is almost certain that component will be provided with a new seal when we receive it from the supplier. So as long as we are happy with the 5-year period, then we don't need to do anything further for the seal on the heating element.

If, for example, we felt that the heating element should be replaced in 5 years and the seal of the element in 4 years, normally we will combine the two replacements into one Planned Maintenance schedule, and use the shortest frequency, 4 years. Where parts cost only a few dollars it is quite safe to make these sorts of changes, it saves a lot of costs in Technician's hours. This step is called PACKAGING of maintenance tasks and is used to combine the tasks into a maintenance program that minimizes the number of separate interventions where it does not affect the integrity of the maintenance strategy.

The remaining seal, (Component 12), can be replaced on Time Based Replacement, again we might estimate 5 years.

(I am well aware that I am taking some wild guesses here on the lifetime of seals. When we built the HACS we faced a similar problem, so we dismantled a great number of valves in the brewery and inspected the seals very carefully for wear and perishing, in order to give us some baseline on the life of the seals depending on their environment. Refer to section 8.5)

9.6 THE MAINTENANCE PLAN

Summarising the above, we now have a basic maintenance plan from the FMEA:

FAILURE MODE	MSI	MAINTENANCE STRATEGY	Frequency
Heating element burned out	Heating element	Time based replacement	5yrs
Fuse blown	Fuse	Run to Failure	n/a
Thermostat not connecting	Thermostat	Time based replacement	5yrs
Cable burned	Power Cable	Time based replacement	3yrs
Thermostat early disconnection	Thermostat	Covered above	n/a
Thermostat fails to disconnect	Thermostat	Covered above	n/a
Cracked kettle	Kettle body	Inspection	1yr
Perished seals	Level gauge seal	Time based replacement	5yrs

CILT Activity:

CILT	Kettle cleaning	Remove limescale by boiling with vinegar	6 months

So, to maintain our kettle in good operating condition indefinitely, how many Planned Maintenance schedules and a CILT activities do we need?

Well actually we only need to write 3 Planned Maintenance schedules!!!

Because the frequencies are the same, we can package and build our Planned Maintenance schedules as:

- Replacement: Heating Element, thermostat and level indicator seal, every 5 years.
- Replacement: Power cable every 3 years.
- Inspect for cracks once a year.

Plus, a CILT of cleaning with Vinegar every 6 months, and we could include the crack inspection in that as it is very quick to do.

We have a Maintenance strategy, but to be clear this is not yet a Planned Maintenance Schedule. A Planned Maintenance schedule will mention the frequency, resources, job type, time required etc.. and will be accompanied by a JOBPLAN, step by step instructions as to how to carry out the task. Where necessary the JOBPLAN may have an attached SOP to show how the task is carried out when it is complex.

Converting the maintenance strategy into Planned Maintenance schedules, Job Plans and SOPs is described in detail in Chapter 13.

Failure Mode Effect Analysis is only the first step in building a Planned Maintenance system. It must never become the whole direction or effort of the Planned Maintenance system development (which is the impression that several authors give), it is one of many important activities, but it has no value if it does not lead to the development of Planned Maintenance schedules and Job Plans.

CHAPTER TEN: 5-WHY ROOT CAUSE FAILURE ANALYSIS

10.1: CONVENTIONAL 5-WHY PROCESS

Root Cause Failure Analysis (RCFA) is also known as Breakdown Analysis (BDA) or 5-Why Analysis.

RCFA is used AFTER a failure has happened and been repaired (via corrective action) to try to determine the root cause of the failure. Once this is determined we can write a Planned Maintenance schedule to prevent the failure from occurring again, or for a man/method cause we can carry out training or develop an SOP or OPL (Standard Operating Procedure or One Point Lesson).

It's important to understand that corrective action alone does not usually prevent the failure from occurring again. For example, what do you do if your car tyre has a puncture because the tread is so worn that it has little puncture resistance? You can repair the tyre and refit it, but it will surely fail again in a short time. You can replace it with a new tyre, but this is still a corrective action, because it will eventually wear and puncture again. The preventive action would be to introduce an inspection, to measure the tyre tread, for example every 6 months, and to replace the tyre when the tread is less than 2mm depth.

Whereas FMEA is a proactive technique (identifies failures that have not yet occurred), RCFA is a reactive one (Identifies the root cause of failures that have occurred and have been repaired).

RCFA is very much a bottom-up approach to implementing a Planned Maintenance system, one failure at a time.

If every failure that occurs is diligently analysed and the root cause effectively eliminated, then eventually we could reach a point of very low unplanned downtime by addressing each failure one by one, using RCFA. However, this is like correcting aircraft maintenance failures one crash at a time, it is a slow process, it's unstructured and wastes a lot of resources.

In a brewery with a high rate of unplanned downtime, say 30%, this will mean that there are several hours of breakdowns per day in each area (not all unplanned downtime is breakdowns), which could easily be between 4 and 10 different breakdowns. The reality is that if this is the amount of unplanned downtime, then there is so much "firefighting" going on (corrective repairs) that there is not the time available to carry out several good RCFA analyses every day.

Also, RCFA may identify a failure in a component, and we may produce a Planned Maintenance schedule to prevent the failure, but it does not mean that the component is then included in a structured hierarchy. We end up with a random list of MSIs in the CMMS with absolutely no method of classification, so that what has been developed cannot easily be copied to another machine or another brewery.

Ideally we should apply a structured methodology to build a Planned Maintenance system, as was done in the HACS, and then we apply RCFA to any remaining breakdowns that occur.

The fact that we have RCFA is also a reason why we don't need to complete a perfect FMEA and identify every possible failure mode and eliminate it with a Planned Maintenance schedule. As stated before, we are not a high-risk industry, and if we carry out a process that eliminates over 90% of breakdowns, we can catch the remainder as they occur using RCFA in our daily operations.

This is a very important point that allows us to effectively reduce the time we put into Planned Maintenance system development, and if we can strike the right balance, as I think we did with the HACS, then we can build an effective Planned Maintenance system with minimum resources.

If you refer back to the supportability project described in Chapter 5, that approach identified 577 MSIs when FMEA was rigidly applied by the manufacturer, but only 82 when the same machine was later analysed using the ways of working we developed for the HACS.

If we missed a few in the HACS, it's OK, we will catch them with RCFA/BDA and then use that finding to update the HACS.

Breakdown Analysis/RCFA can only be carried out if the equipment has been repaired and is operational again following the breakdown. If the problem is not fixed, such as a recurring process deviation, then we need to use other tools such as Ishikawa (Fish Bone Analysis) to eliminate possible causes.

A Breakdown Analysis sheet is shown in Figure 48.

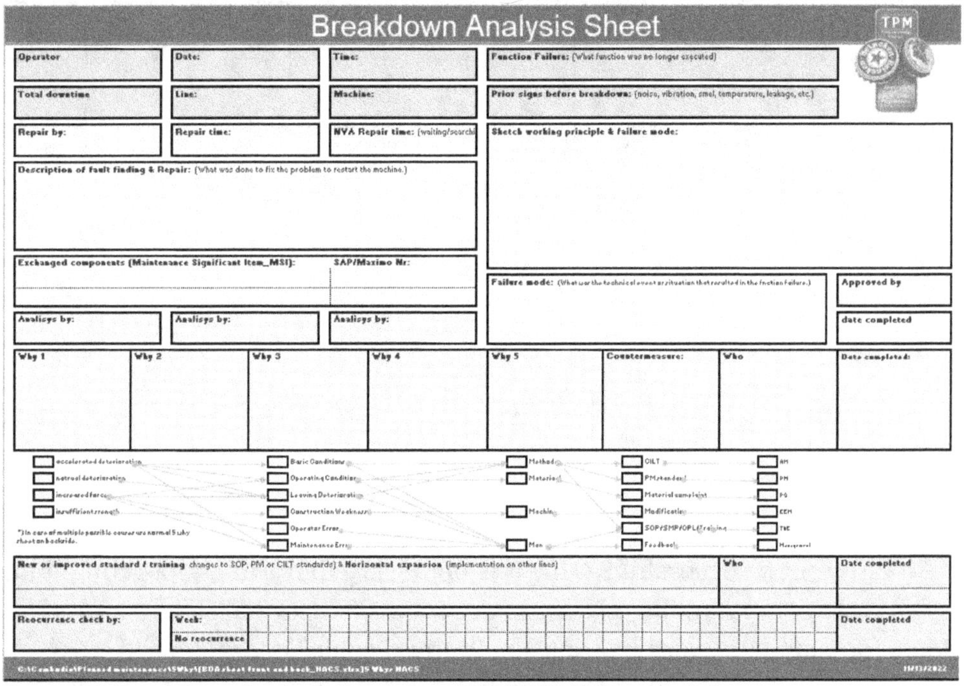

Figure 48: Breakdown Analysis Sheet, from HEINEKEN TPM materials.

The Breakdown Analysis Sheet records the time and duration of the Breakdown, a description of the fault-finding and repair carried out, the components replaced and then there should be a description of the functional failure, the failure mode and a sketch of the relevant working principle.

Ideally the top section is filled in by the operator and the central section by the technician who did the repair.

The third section is the failure analysis section, and this analysis is usually done at a later time, though I much prefer it to be done on the line as part of the breakdown whenever possible. Sometimes further information is needed, like checking if a lubrication schedule was completed, so then it cannot be filled in straight away.

The analysis should always be completed by a cross functional team, for example to be carried out by one person from Engineering, one from Production and one from the Quality team.

There is space to summarise the 5-Why process on the front page of the sheet, on the back of the sheet there is more space to allow for a more detailed working of the 5-Why (Figure 49).

Failure Mode	Potential causes										ert	Actions			
	Why (1)	Check	Why (2)	Check	Why (3)	Check	Why (4)	Check	Why (5)	Check		CORRECTIVE ACTION	Check	PREVENTIVE ACTION	Check

Figure 49: Back of BDA sheet for 5-Why working, from TPM materials.

To carry out a successful 5-Why it is necessary that the people completing it have a clear understanding of how to describe a Functional Failure and a Failure Mode, for example "worn bearing" or "burned cable" or "faulty sensor".

For this failure mode (or some failure description let's say it is WORN BEARING), we would then ask the first "WHY (1)?"

Some people like to write every possible WHY? that they can think of in the WHY? column, so in the first WHY (1)? column for the failure mode WORN BEARING we could write:

- No grease.
- Too much grease.
- Incorrect grease.
- Old age.
- Dirt/contamination.

But I find that this is usually unnecessary, because the whole point of the 5-Why is that we already have repaired the failure and we should have the broken part literally in our hands. We don't need to speculate on possible failure causes if we can see it in front of us. We only need to speculate if we don't know, which is when we would use the back sheet and consider various possible causes in order to investigate further. We would use the back sheet more frequently for electrical failures when we are often not sure of the actual cause, such as for some of the failures described in Chapter 7.

But if the bearing is dry and worn out, then we can be sure that in WHY (1)? we only need to have, "NO GREASE"- and we don't have to list all the other possible WHYs.

When we look and see that there is no grease, we next need to investigate why that occurred, (or why lubrication didn't occur) to answer the 2nd WHY? Again, we can speculate that there are several possibilities, such as we forgot to grease it, or it is not on the greasing schedule, or the grease nozzle is blocked, but again we can investigate, and we might find that it is not on the greasing schedule. So, no further speculation is needed, the answer to WHY (2)? is NOT INCLUDED IN GREASING SCHEDULE.

In this case we can stop after only 2 WHYs? Actually 2 or 3 WHY's? is often sufficient, it is rare that we need all 5.

We could ask WHY (3)? is it not in the greasing schedule. It could be that the technician who made the schedule did not know the machine very well, or that the schedule from the manufacturer was faulty.

But clearly if this bearing is missing from the greasing schedule, we now need to change our METHOD. The corrective action will be to grease the new bearing, the PREVENTIVE action will be to add the bearing to the greasing schedule and to make sure that whoever does the greasing is informed and trained.

To be clear, the CORRECTIVE action describes what you repaired or replaced, the PREVENTIVE action should describe the maintenance actions that you implemented to prevent reoccurrence of the failure.

Multiple possible causes do not need to be listed under each WHY? if you have examined the part and have a good idea of what happened.

For many breakdowns the 5-why may show that the root cause is a failure of the MAN (or woman) such as the operator did not follow an SOP, so we will need to retrain and reinforce the importance of the SOP. Or perhaps it is the METHOD, the SOP was missing or incorrect, and then we just need to update the SOP, such as in the example that the bearing is not included on the greasing schedule. Or it may be MATERIAL, such as empty cans that are out of specification, and then we are going to reject that material and take it up with the supplier.

But when the root cause is a MACHINE (component/MSI) issue, we need to look a bit further at the true failure cause, which is the topic of the next section.

(In many HEINEKEN breweries the BDA/5-Why is now completed in the One2Improve system, a digital tool that is used for recording maintenance tags (Maintenance requests for Corrective Maintenance or Safety), completing 5-Whys, carrying out audits, linking to Maximo and other functions such as CILT checks. One2Improve is a fantastic tool, but it is beyond the scope of this book.)

10.2: UPDATED 5-WHY FOR THE HACS

I have been very fortunate to work with an Engineer in Cambodia by the name of Akshay Setlur Ramamohan, who I have coached as a trainee for some years and who has become a strong supporter of the HACS system. Akshay has been the one leading the Engineering team in Cambodia to implement the HACS and helped to deliver the cost savings and reduction in unplanned downtime that have proven the HACS to be the right Planned Maintenance system for us. Akshay is highly energetic and always coming up with new ideas that help us to improve the HACS and our reliability performance in Cambodia.

One of Akshay's ideas was to modify the 5-Why failure analysis tree to take into account the implementation of the HACS system.

The failure analysis tree is a structured way of trying to drill down into the root cause in the case of a component/MSI failure (MACHINE cause) and guide the Technician in finding the true root cause and best corrective actions.

I will now explain it in some detail, bold capitals indicate the questions in the steps of the analysis tree.

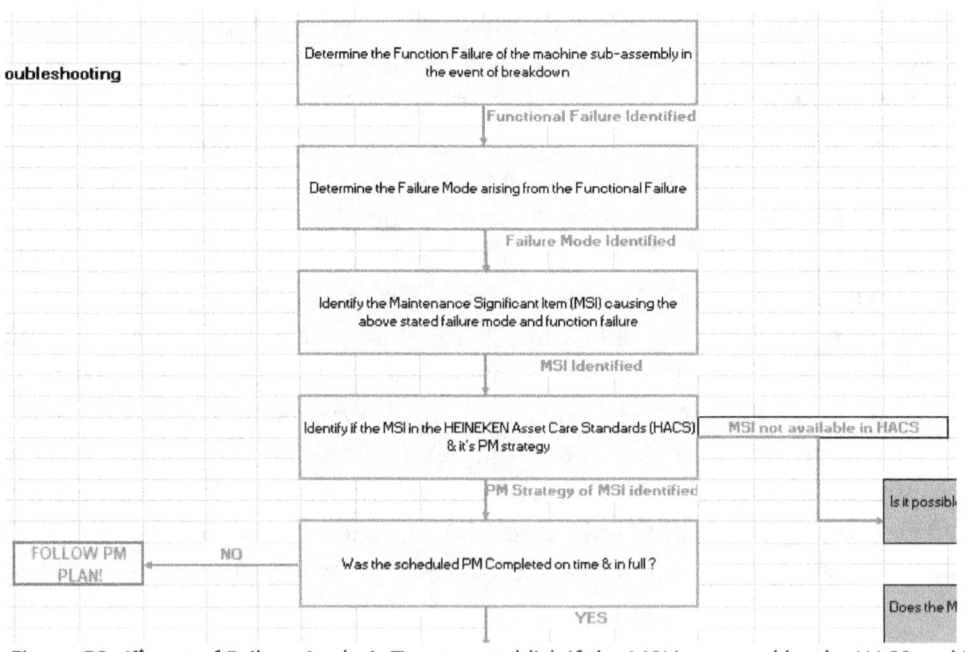

Figure 50: 1st part of Failure Analysis Tree to establish if the MSI is covered by the HACS and if the Planned Maintenance schedule was followed, from updated Failure Analysis tree.

The first part of the flow chart is shown in Figure 50 and it serves to identify the functional failure, failure mode and MSI, and then to find out if the MSI that has failed is included in the HACS system and has a Planned Maintenance strategy : "IDENTIFY IF THE MSI IS IN THE HEINEKEN ASSET CARE STANDARDS (HACS) & IT'S PM STRATEGY".

If the answer is NO: "MSI NOT AVAILABLE IN HACS", then it is a gap in the HACS and we go to section 2 on the right of the chart, which follows the Maintenance Task decision flowchart reviewed earlier as per Figure 47. At this point in our fault-finding, we know that as the MSI was not covered by the HACS, and that the correct step is to update the HACS. This action will help to improve the quality of the HACS over time.

But if it was covered by the HACS, the next question is "WAS THE SCHEDULED PM COMPLETED ON TIME & IN FULL"? If NO, then the answer is equally simple, follow the schedule (we might want to investigate why it was not followed, such as if there is too great a backlog in Planned Maintenance tasks).

But if the HACS schedule was followed and the MSI still failed, then we need to find out why, why is it that the HACS failed to prevent the breakdown?

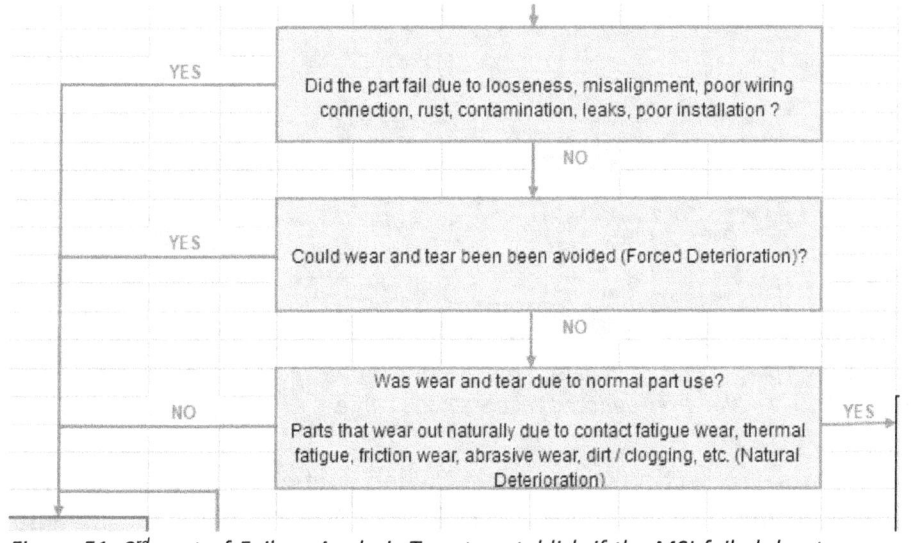

Figure 51: 3rd part of Failure Analysis Tree to establish if the MSI failed due to normal use, from updated Failure Analysis tree.

As shown in Figure 51, we next try to establish if the failure was due to the MSI reaching the end of its designed life under normal conditions, or if the failure was in some way accelerated. So first we ask if it is misaligned, rusted, contaminated etc. and if not if the wear could have been avoided. But if the wear and tear is judged as due to normal use, we will move right to section 5 (Figure 53), if NOT, we move down to section 4 (Figure 52).

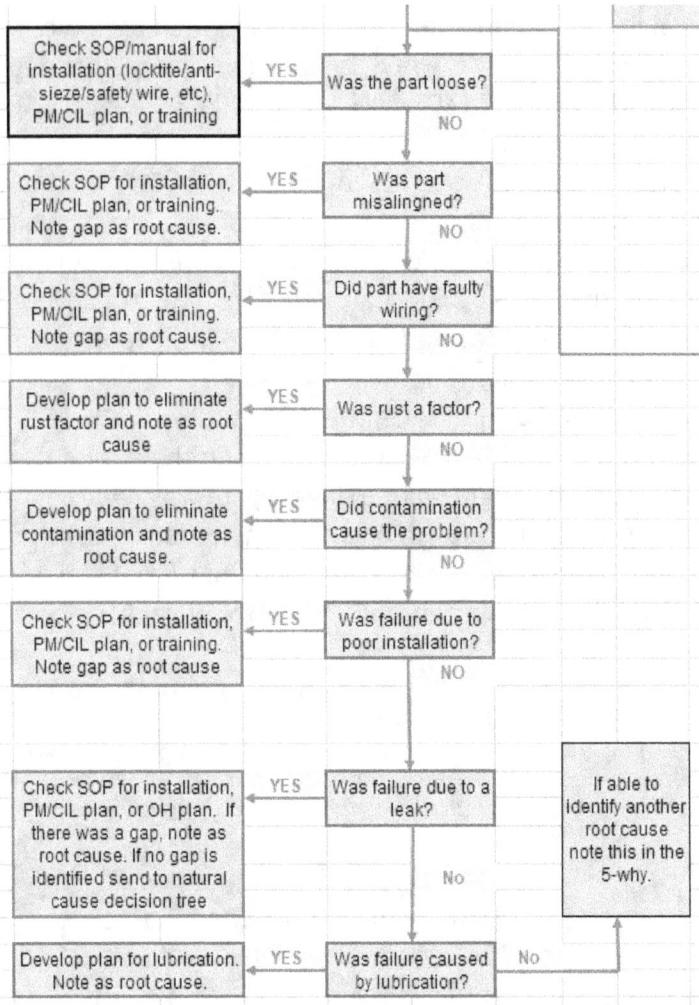

Figure 52: 4th part of Failure Analysis Tree to establish the reason for accelerated wear and suggest preventive action from updated Failure Analysis tree.

We will move to Section 4 in Figure 52 when we believe that the failure was NOT due to normal use/wear, we suspect that the MSI failed in some accelerated or unexpected manner. So, following the questions we check if it was loose, misaligned, corroded, contaminated, incorrectly installed or failed due to inadequate lubrication.

If any of the above is the case, then the appropriate action is specified, such as checking or developing an SOP for installation or making a lubrication plan. One occurrence we have from time to time is loose bolts, and this we normally address by specifying to use "Loctite" on the thread of that bolt, which prevents a bolt from loosening under vibration.

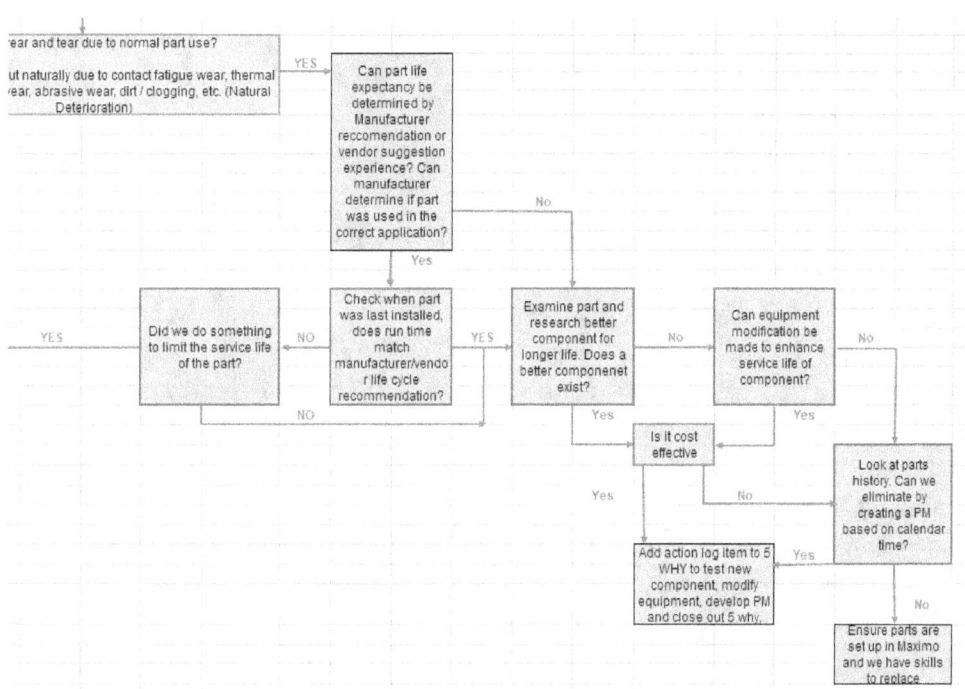

Figure 53: 5th part of Failure Analysis Tree to establish normal service life of the MSI and possible alternatives from updated Failure Analysis tree.

We only go to the very last section of the chart, as shown in Figure 53, if we find that the MSI failed naturally as a part of its normal service life.

In this last part of the flowchart, we first check if the part reached the end of its normal service life as specified by the manufacturer. If NOT then we return to section 4 to analyse why the actual life was shorter than expected.

But if the part did follow the expected lifetime, or we don't know the expected lifetime, then we ask if a better component exists that is a higher quality and is cost effective. If so we can change to using that component. Often it happens that our suppliers change a part to an improved specification in this regard. One example is that we might change a bearing with manual greasing for a bearing that is sealed for life, if appropriate.

We also ask if a modification to the equipment can enhance the service life of the component. If this is the case, then we also check if that is cost effective before proceeding.

But if there is no better-quality component available (or the cost is not justified), and the equipment cannot be modified, the last question is to see if there is a calendar time aspect to the failure, based on replacement history: "LOOK AT PARTS HISTORY. CAN WE ELIMINATE BY CREATING A PM BASED ON

CALENDAR TIME". This last question is a re-check of the logic in the HACS, to see if any data on the lifetime of the component indicates a time to failure that was perhaps missed in the HACS or incorrect. So, we ask if we can eliminate the failure by creating a Planned Maintenance schedule based on calendar time (Time Based Replacement)?

If none of the above applies we will continue with the part as is with the Planned Maintenance schedule unchanged.

Consider as an example a proximity sensor that has failed, and the failure is in the internal electronics with no sign of water damage or other damage. We find that a Planned Maintenance schedule is included in the HACS to replace the sensor every 10 years, but it failed after only 3 years. So, it is in the HACS, and we did follow the HACS. We cannot assign any cause in section 4, there is no indication that it was loose, misaligned, corroded, contaminated, incorrectly installed or failed due to inadequate lubrication. The expected life from the manufacturer is 10 years.

We can decide whether to research an alternative component that could have a longer life. Perhaps this type of sensor is of poor quality, and we decide to specify a better-quality manufacturer. Assuming we cannot modify the machine, or it will not help, we ask in section 5 if we can create a Planned Maintenance schedule based on calendar time. So, we should next look at how many sensors of the same type failed and after what lifetime.

If this event is unique, then we will replace it and see if it happens again. But if we find other sensors fail also after a few years, then we will be needing to change to a Time-Based Replacement, or upgrade to a more durable model.

The analysis developed by Akshay is carefully designed to check for every possible assignable failure cause and to only move to what is effectively a conclusion of "unknown" after exhaustive checking. Even in that case we try to check if this is a regular failure pattern for this component, or if it is truly a one-off event.

CHAPTER ELEVEN: CMMS & MAXIMO

11.1 CMMS HISTORY

Nolan and Heap would have gathered and analysed their data without having much help from electronic calculators, an electronic database or word processors to help write their report. The first basic Word Processors appeared in the 1970's, as well as the first computerisation of maintenance tasks, so it is highly unlikely that they had much electronic assistance.

All of their data probably had to be manually collected and arranged, and their 480-page report was surely manually typed by a pool of typists making the whole process highly laborious.

John Moubray first published his RCMII in 1991, by which time the IBM PC with WORD was available, so he may have had an easier task.

Today we can rapidly analyse data, and as I write this book I can instantly and easily review and change the document using Microsoft Word.

Neither Nolan and Heap nor Moubray mention a Computerised Maintenance Management System (CMMS) in their publications, but they take some time to explain manual maintenance planning and control systems including maintenance checklists and the use of various information worksheets.

Smith and Mobley in their much later "Rules of Thumb for Maintenance and Reliability Engineers", (Mobley, 2008, p. 60) only mention CMMS in passing and do not focus at all on its pivotal role in building an effective Planned Maintenance System.

The earliest versions of CMMS appeared in the 1960s and used punch cards and mainframe computers to update computerized records and track maintenance tasks. In the 1970's, punch cards gave way to checklists fed into CMMS systems by Technicians at the end of their shifts (What is a cmms, 2013) https://www.ibm.com/topics/what-is-a-cmms

CMMS gained greater usage in the 1980's and 90's as computers became smaller and more affordable, and in the 2000's the development of the internet allowed for information sharing across networks. Some CMMS packages are now completely cloud based.

11.2 DEFINITION OF CMMS

A **Computerized Maintenance Management System** (CMMS) is a software package that centralizes maintenance information about assets in a database and organises the information in a structured way. It manages the planned maintenance schedules (work orders) for the assets and applies scheduling and reporting functions to them, it manages key spare parts information such as parts usage and cost of spares, and finally will have reporting and analysis functions.

CMMS systems are found in aerospace, manufacturing, oil and gas production, power generation, construction, transportation and other industries.

There are many different CMMS packages available, some are designed for specific fields such as fleet maintenance.

The core of a CMMS is it's **database** that organizes the information about the assets of the company, as well as the spare parts, materials, and other necessary resources.

The CMMS should provide the following capabilities:

Asset registry: Store, access and share asset information such as:

- Manufacturer, model, serial number and equipment class and type.
- Associated costs and codes.

- Location and position.
- Performance and downtime statistics.
- Associated documentation such as supplier repair manuals and SOPs.

Work order management: The main function of CMMS, work order management includes information such as:

- Work order number.
- Description and priority.
- Order type (inspect, replace, calibrate, lubricate).
- Personnel assigned and spare parts used.

Work order management should also include capabilities to:

- Automate work order generation.
- Reserve materials and equipment.
- Schedule and assign employees, crews and shifts.
- Review status and track downtime.
- Record planned and actual costs.

Preventive maintenance: Automate work order initiation based on time, usage or triggered events. Triggered events might include an input from a Condition Monitoring system indicating excessive vibration, or from a process control system such as Brewmax that is able to record the number of operations of a valve that we know should be maintained after so many operations.

(Maximo works well in automating work order initiation based on time, we are implementing automation of work orders from the triggered events given as examples above).

Spare parts inventory management: Manage inventory of spare parts and materials across storage areas and facilities. Track spare parts costs.

Reporting, analysis and auditing: Generate reports across maintenance categories such as asset availability, materials usage, labor and material costs and supplier assessments. Analyze information to understand asset availability, performance trends, MRO inventory optimization and other information to support business decisions and gather and organize information for audits.

11.3 ASSET HIERARCHY IN MAXIMO:

The Asset Hierarchy is the way that we organize the site, departments, assets, assemblies, sub-assemblies and components in a structured way using parent-child relationships so that every single component in the database can ultimately be linked back to a higher-level sub assembly, assembly, asset, department and site.

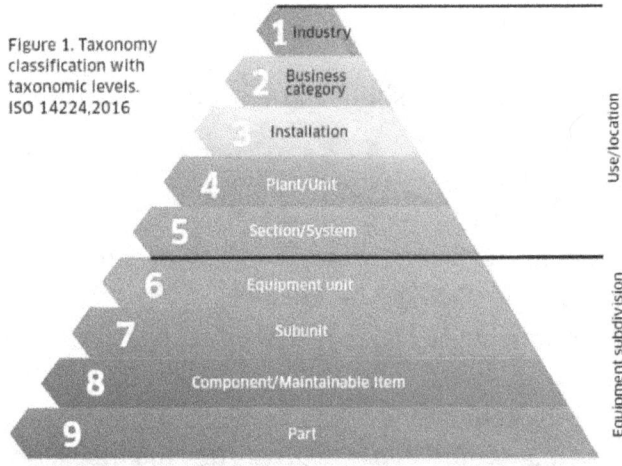

Figure 1. Taxonomy classification with taxonomic levels. ISO 14224.2016

1 Industry
2 Business category
3 Installation
4 Plant/Unit
5 Section/System
6 Equipment unit
7 Subunit
8 Component/Maintainable Item
9 Part

Use/location

Equipment subdivision

Figure 54: Hierarchical structure from ISO 14224, 2016

Such a hierarchy means that every component has a parent line that can be followed. ISO 14224 in Figure 54 is the most commonly used asset hierarchy standard.

An accurate asset hierarchy is of paramount importance in managing and building a Planned Maintenance System.

- It allows you to aggregate and collate maintenance work orders by asset or assembly.
    - This then allows for scheduling maintenance of all relevant assemblies, sub-assemblies or MSIs in an asset when that asset is off-line.
    - This also allows for collation of cost data by asset, or if needed by assembly or MSI.
- It allows for easy identification of MSIs that need maintenance vs BOM items that do not need maintenance.

- It helps to ensure that all MSIs have been included in the Planned Maintenance system. The HACS contains over 10 000 Planned Maintenance schedules which need to be managed in a typical brewery.
- It helps you to standardise naming conventions: (For example, in a brewery there are hundreds of gearboxes and motors. To be able to find them in different assets, check for spares and share Planned Maintenance schedules, we have to always use the same name. We settled on Motor-Gearbox, and we always include them as a set. But we could have used electric motor, 3 phase motor, conveyor motor or many other different descriptions that would make database management very difficult if we changed the description on different assets).
- The hierarchy allows you to manage your resources: by having an asset broken down into sections, you can schedule maintenance on different sections as sufficient resources are available.
- It is easy to copy the Planned Maintenance schedules from one site to another for assets that are the same or similar, and the hierarchical structure allows you to check for any small differences, such as a slightly different configuration.
- When you are developing maintenance schedules for a machine, if you have a good hierarchical structure for a similar machine, then it is easy to copy the schedules for common or similar assemblies and sub-assemblies. This has allowed us to accelerate development of the HACS. For example, if we have a HACS file of Planned Maintenance schedules for a Sidel can filler, and we wish to build the Planned Maintenance schedules for a Krones can filler, then if we have a clear hierarchy for both it becomes very easy to see the similar assemblies and sub-assemblies and then to copy the Planned Maintenance schedules where appropriate.

The above 2 points have allowed us to build an end-to-end planned maintenance system in 3-4 years with limited resources, that otherwise would have taken several times more resources.

In the same vein, the above is allowing us to expand the HACS globally at a high speed.

THE MAXIMO HIERARCHY

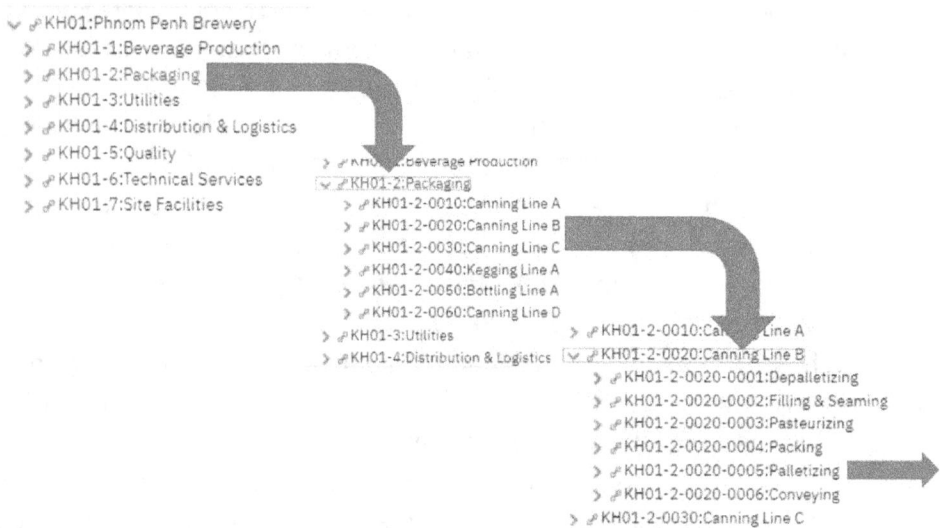

Figure 55: Maximo Hierarchy from Maximo database.

Figure 55 shows part of the Maximo Hierarchy for our brewery in Cambodia. By way of example, we can drill down from the BREWERY (KH01) to the PACKAGING department (2) to CANNING LINE B (0020) and then to the PALLETISER (0005), machine KH01-2-0020-0005.

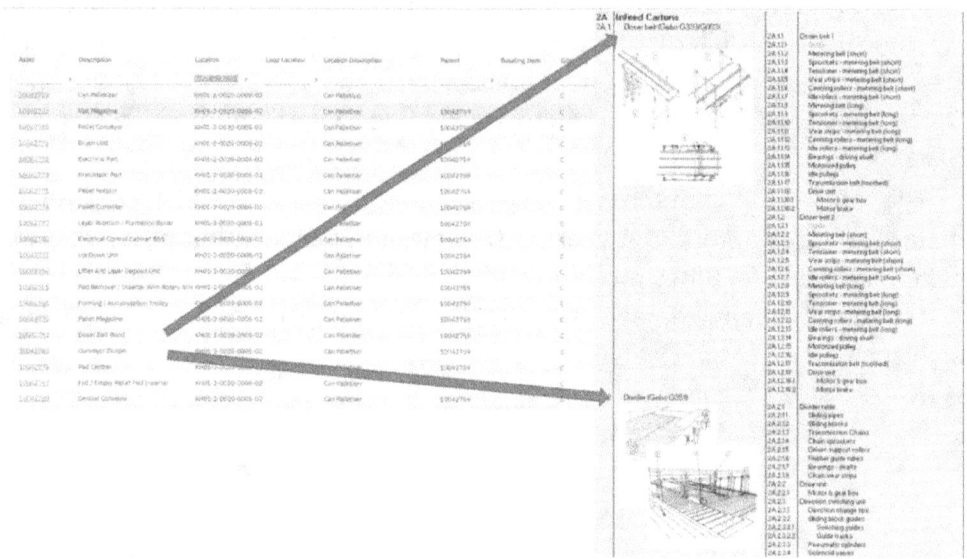

Figure 56: Machine Hierarchy in Maximo from Maximo database.

140

Figure 56 shows the Palletiser has asset number 10042759, and the assets underneath are belonging to the parent 10042759. These assets are the construction groups in the hierarchy that is explained fully in the next chapter.

It leads to chaos if you try to build a Planned Maintenance system without structured asset numbering having first been implemented, as without this you have no way to organize the assets and develop their Planned Maintenance schedules.

The full asset number for the palletiser in this example is KH01-2-0020-0005-10042759. In Chapter 15 I have described the need to have on-site an engraving machine to facilitate the labeling of the assets with the correct asset number.

11.4 MAXIMO UPLOAD SHEETS

I will describe in the next chapter how the HACS files are built and how they reflect the hierarchy inherent in Maximo. As this chapter is about CMMS, I will describe here how the output of the HACS files are loaded into the CMMS.

In order to upload the Planned Maintenance data into Maximo, the data has to be in the correct format, and there are three different sheets required for this:

- Asset upload sheet: Describes the assets and the hierarchy (Figure 57).
- PM upload sheet: Describes the Planned Maintenance schedules (Figure 58).
- Job Plan upload sheet: Describes the step-by-step Job Plan of the Planned Maintenance schedule (Figure 59).

THE ASSET STRUCTURE UPLOAD SHEET:
The asset structure with parent-child relations is the backbone of any asset management system. Planned Maintenance plans are linked to the assets, assemblies, sub-assemblies and MSIs through this sheet:

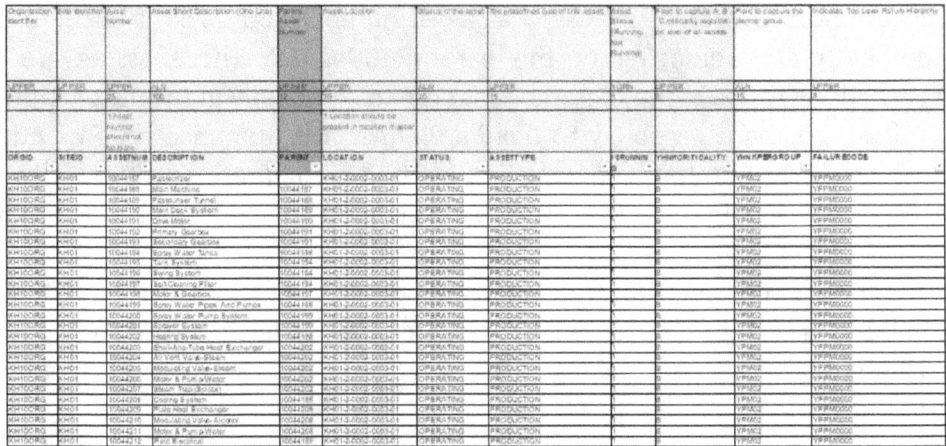

Figure 57: Maximo Asset upload sheet from HACS database.

THE PLANNED MAINTENANCE PLANS UPLOAD SHEET

The Planned Maintenance plan (Figure 58) sets the link between the asset and the Planned Maintenance schedules. Tasks in Planned Maintenance schedules shall be executed according to a certain frequency.

A – An asset can have multiple Planned Maintenance schedules.

B – Job Plans can be linked to multiple assets and multiple Planned Maintenance plans.

C – A Planned Maintenance schedule has a frequency.

D – To start the first cycle of a Planned Maintenance schedule a "next due date" must be set.

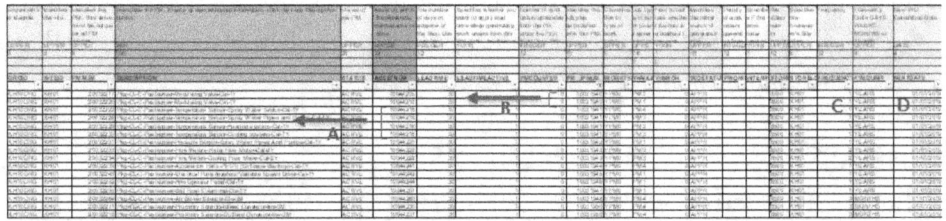

Figure 58: Maximo Planned Maintenance schedules upload sheet from HACS database.

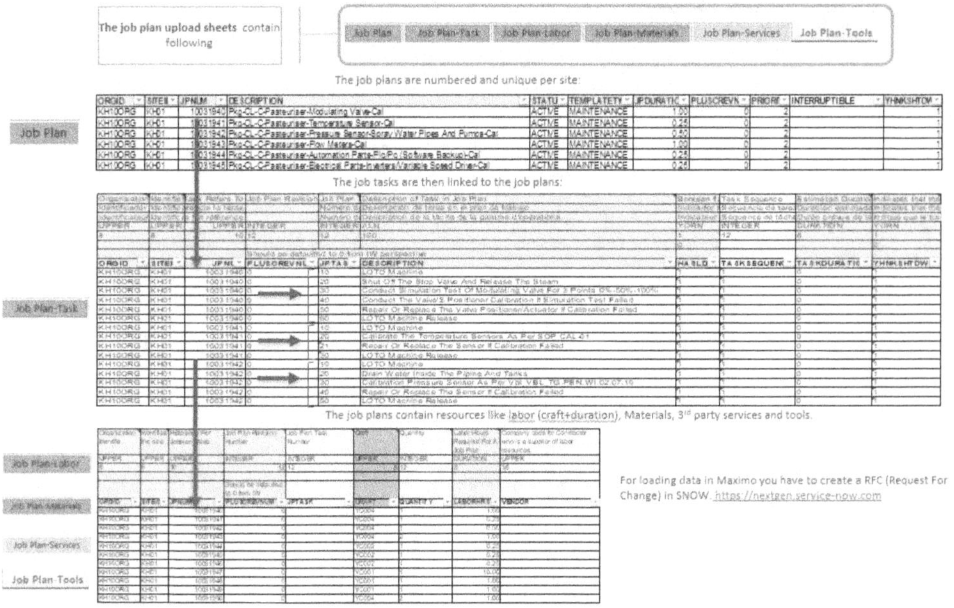

Figure 59: Maximo Job Plan upload sheet from HACS database.

The Job Plan upload sheet (Figure 59) contains the Job Plans, including the labour, tasks and tools required. This sheet links the Job Plans to the Planned Maintenance schedules. It is the Job Plan that is printed for the Technician as an instruction to carry out the required Planned Maintenance schedule.

The three sheets have to be completed for each machine (main asset) and when the global web-based version of Maximo is in use, the sheets have to be sent to IBM for upload.

The HACS COMPLETE file that we use in Excel to develop the machine hierarchy and the Planned Maintenance schedules and Job Plans, was deliberately designed to have the same column structure as the Maximo upload sheets, which makes the data transfer to the upload sheets much easier to carry out.

Still there is a huge amount of work required in collecting and compiling all the data for the upload sheets. There is some similarity here to the worksheets described by Nowlan & Heap and by Moubray, and it is necessary to have the clerical competencies in place to complete and update the upload sheets. But, once they are in the system, they are easily modified and updated and become fully automated unlike the manual systems used in the past.

Here I have only shown part of each upload sheet, as it is beyond the scope of this book to give detailed instructions on completing each of the upload sheets. For example, the asset structure upload sheet has a total of 200 columns for each asset line entry.

CHAPTER TWELVE: BUILDING THE HACS HIERARCHY

12.1 WHERE TO START?

As described in Chapter 5, after the 2017 strategy meeting with Jan Paul we had decided to start on the project to "Fix Planned Maintenance in Asia" by first building a Planned Maintenance system for the Gebo/Sidel SCALE canning lines, of which we had 9 in the region at that time.

One of my first steps was to look at the output from the supportability project, as described in Chapter 5. It was clear that this analysis was unstructured (no hierarchy), and did not result in any Planned Maintenance schedules, so it was of no use for what I planned to do. I already knew that following a structure, as is inherent in the CMMS, was absolutely essential.

I also obtained many of the existing Planned Maintenance schedules that we had in place around the world, including from Brazil, Nigeria, Holland, Papua New Guinea, the UK and Russia. Whilst it was clear that a lot of Engineers around the world had tried their best with limited resources to produce Planned Maintenance schedules, none of them had developed a systematic and thorough Planned Maintenance system.

Most of the schedules did not follow a Hierarchical structure, often MSIs would just appear with no connection to an assembly or an asset. Most of the maintenance files were not complete in that the whole machine was not covered. Most did not have specific standards for Inspections, instead using subjective instructions such as "check the motor".

All of the schedules were in different formats. In one location there was a list of maintenance tasks, built like a checklist, in another we have an excel table listing Planned Maintenance tasks, but with no hierarchy. In a third brewery inspections are in a Word document format with lots of photos and figures. None of the formats followed the structure of the Maximo upload sheets described in the previous chapter, so transposing them into the Maximo system would be a very challenging task.

In order to reduce the workload of implementing Planned Maintenance by sharing Planned Maintenance schedules for similar machines, **we need to use a common format**.

It was also well known from my day-to-day discussions on the shop floor with Engineers that what was available from all of our equipment suppliers in terms of Planned Maintenance schedules was completely inadequate and sometimes almost non-existent from an RCM perspective. Some manufacturers do provide limited maintenance instructions, mostly these refer to lubrication tasks, but occasionally there are some overhaul tasks as well. In general, they follow the philosophy that *"every component has a certain lifetime after which it must be overhauled, inspected and if necessary replaced"*, which we now know is incorrect. Where they specified Inspection tasks there was usually no standard.

12.2 EXISTING PLANNED MAINTENANCE SCHEDULES IN THE SCALE LINES

Most of the SCALE canning lines in the APAC region were running with very low levels of Unplanned Downtime (UPD), so this might be used as an argument that a structured and comprehensive RCM based Planned Maintenance system is not needed (downtime was much higher on the bottling lines, Brewhouses and non-Scale canning lines).

I decided to examine and compare the Planned Maintenance systems of some of the SCALE canning lines, so I visited several of the breweries in Vietnam as well as Cambodia and Singapore to examine their Planned Maintenance systems in operation.

The table below shows a part of the hierarchy of the Palletiser (more about building the hierarchy later in this chapter), and what Planned Maintenance schedules were available for the identical canning lines for those MSIs in this section of the machine. I have deliberately left off the names of the specific breweries and do note that this is only a small part of the complete hierarchy that I checked, using the palletiser as a baseline.

Figure 60: Comparison of Planned Maintenance schedules available from the supplier and developed for different SCALE canning lines for one section of the palletiser.

In Figure 60 the first empty column on the left shows that Planned Maintenance schedules from the supplier for this part of the machine do not exist. The next columns are for some of the SCALE lines in Cambodia, Vietnam and Singapore showing that there are completely different amounts of coverage of Planned Maintenance schedules at the different sites for the same MSIs. Only one line covers more than 50% of the MSIs with Planned Maintenance schedules, the rest have a lot missing, shown as red boxes.

It is clear that even with SCALE canning lines (which are all exactly the same) that there is effectively no sharing of the Planned Maintenance schedules, even within one country.

Then I examined the quality of the Planned Maintenance schedules that did exist, just like the schedules I had received from around the world, they were not specific. There were no well-structured Job Plans, there was no structured hierarchy and there were no standards for the inspections.

Smith (Smith, 1993, p. 6) identifies Planned Maintenance variability between similar units as a common maintenance problem (He is the only RCM author I found who does so) and suggests it derives from feelings at the plant level organization of "I'm not like them" and "we know what's best for us" which I have seen in our breweries and seems to be a universal human trait.

I even discovered in one large brewery, the **Planned Maintenance** schedules were not being issued from Maximo but were kept in an Excel file for manual scheduling, so that most of the benefits of a digital CMMS were lost.

I carried out hands-on audits on the lines with the Technicians to try to understand how the maintenance was being carried out, I recall looking at the four palletiser hoist motors with a Technician in one location and asking him what Planned Maintenance schedules he had for these. He answered that there was not one, but *"don't worry sir, I know I have to check them every 6 months and I won't forget."*

And it was then that I realized how some breweries were managing to achieve low downtime with an inadequate Planned Maintenance system, it was because they had good Technicians who knew the machines.

But what happens when those Technicians leave or retire, or we keep expanding our operations and reducing the pool of skilled Technicians. How do we pass the knowledge on to new staff? How do we sustain the performance ?

What was now clear was that to build a Planned Maintenance system for the SCALE canning lines I needed to start from scratch.

12.3 FACILITATED WORKING GROUP APPROACH

Regan describes a facilitated working group approach for RCM implementation (Regan, 2012, p. 31), and describes the benefits of such an approach in terms of utilizing the knowledge of Operators and Technicians who know the machine intimately, and this is indeed part of the process we followed.

But what is missed by Regan and others is the need for a strong Steering Committee to ensure that the working groups carrying out FMEA and building the Hierarchy and Planned Maintenance schedules all have a common understanding of the depth of analysis, ways of working and approach to be used.

I joined forces with Nguyen Troung Giang who had been the Brewery Engineer in Hoc Mon brewery in Ho Chi Minh when I was the Supply Chain Director for Vietnam Breweries, so we had already worked closely together, and Kubatbek Alimbekov, then the Planned Maintenance Engineer in Cambodia, a young and smart trainee. They will be referred to as Mr. Giang and Mr. Kuba respectively going forwards from here. I was still in my regional role as Area Supply Chain Director for the APAC region.

Mr. Giang, Mr. Kuba and I formed the Steering Committee, and we developed the standards for building the HACS, including:

STEERING COMMITTEE ROLE

- Assist the working group to source all necessary documentation for the machine, especially P&ID diagrams, design specifications, Spare Parts Catalogues, Maintenance and lubrication manuals, operator manuals.
- Check that the asset numbering has been completed and that there is a structured numbering system in place in the brewery.
- Check that the HACS does not already exist for a similar machine, or that there is not a set of well-developed Planned Maintenance schedules available elsewhere to use as a base.
- Define the common format for the documentation to be produced (THE HACS COMPLETE file), see Section 21.1
- Define a common format for the Machine overview and common numbering of machine sections.
- Define the common format for the Hierarchy including Construction Groups concept.
- Define the common approach to divide assembly and sub-assembly down to MSI level.
- Ensure exclusion of all non-relevant parts in the BOM from the hierarchy.
- Checking and confirming the correctness of the Hierarchy before a working group can start on the Planned Maintenance schedule development.
- Standardise numbering of construction groups, assemblies and sub-assemblies.
- Checking quality of Planned Maintenance schedule development and correct use of Maintenance Strategy flowchart.
- Checking quality of Job Plans and especially standards for inspections.
- Set generic standards for inspections, such as changing wear-strips when 50% worn or chains when 3% stretched.
- Ensuring SOPs are included where needed and are added to Swipeguide (Swipeguide, 2013) and checked.

- Ensure that Job Plans are written at the correct technical level, in that they can be carried out by a qualified Technician who does not have previous knowledge of the asset.
- Checking that for every assembly or component that already has a Planned Maintenance schedule developed elsewhere, to make sure that the existing Planned Maintenance schedule is copied, and the working groups are not creating new Planned Maintenance schedules for the same components when they already exist.

The Steering Committee documented the key points that are described in this section and the next section (Developing the Planned Maintenance schedules so that the standardised approach could be trained and shared more easily to each working group), to ensure that the output of each working group is standardised in terms of all of the points mentioned above.

WORKING GROUP ROLE

The HACS of each asset was produced by a Working Group structured as follows:

- A leader responsible for delivering the completed HACS for that asset. The leader had to have completed my RCM training module.
- Technical experts (Engineering) who know the machine, both mechanically and electrically.
- An Operator who knows the machine.
- Support and guidance from a member of the Steering Committee.

The function of the Working Group is to produce a HACS Complete file for a specific asset according to the guidelines set by the Steering Committee.

We kicked off the Steering Committee with many hours of lengthy discussions about how we would build the Planned Maintenance system, starting with the SCALE canning lines.

The first clear point was that we would start with one machine, so we formed our first working group for that purpose. Because it was the first working group and somewhat experimental, all of the Steering Committee were part of the first working group.

Most texts, and our own TPM system, tell you to carry out a criticality assessment and rate the machines A, B or C, and then start with the "A" machines.

I do not believe in criticality assessments in our industry, because in both the Brewing and Packaging functions, production follows a linear process. In Packaging the cans move from the Depalletiser to the Rinser to the Filler to the Seamer to the Pasteuriser to the Packer and to the Palletiser. If any of these machines has a breakdown, the whole canning line is stopped.

Of course, there can be some off-line machines, like a rejected can compactor that are really not critical, but in general we don't have any machines that we don't need. Of course, criticality can also look at safety impact or effect, but for the canning line we have to make the Planned Maintenance schedules for the whole line, and we cannot choose any one machine above another.

So, we decided to start with the Palletiser because it was accessible, being at the end of the production line, and because mechanically it is quite easy to understand, it is quite an open machine and it's easy to see what is happening. Also, we had studied the palletiser in the evaluation of the Planned Maintenance systems in use in the SCALE lines as mentioned above, so we already had some familiarity with the machine.

Next we had to decide how to break down the first machine (asset) into a Hierarchy.

We started with:

- Standardize the numbering of the machine's main sections (like infeed, outfeed etc..)
- Break the section's down into Construction Groups.
- Do not include the whole BOM, only the MSIs. (So, we have to make a judgement on each component if we think it will require maintenance.)
- Complete the Hierarchy before we start on the Planned Maintenance schedules.

12.4 HACS STANDARD MACHINE SECTIONS

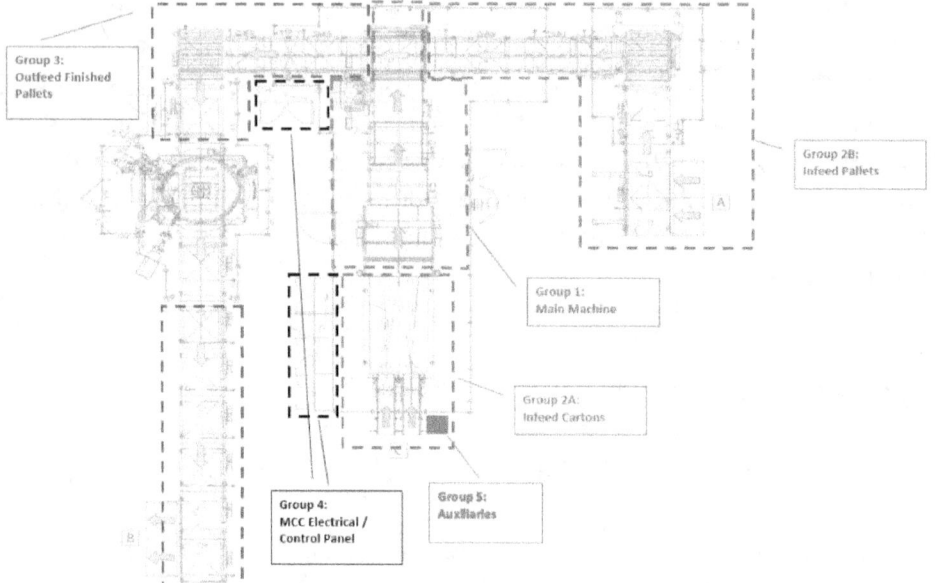

Figure 61: HACS Standardised Machine Sections for a packaging machine, Sidel palletiser HACS file.

We decided on the following standard machine sections:

- Group 1: Main Machine.
- Group 2: Infeed.
- Group 3: Outfeed.
- Group 4: MCC/Electrical control panels.
- Group 5: Auxiliaries.

Where we have more than one infeed, as is the case above with infeed of cartons and infeed of empty pallets into a Palletiser, then we have Infeed 2a and Infeed 2b.

(If you notice that there is a section not included in Figure 61, that part is a stretch wrapper for which we wrote a separate set of HACS schedules).

Figure 61 is the General Arrangement (GA) drawing of the machine, an Engineering layout that is like a birds-eye view. These are usually available in every brewery's Packaging department, so this is a good starting point.

We also standardised on using an Excel file for each machine's HACS standard, and the first main worksheet of the Excel file will be the Standard Machine Sections.

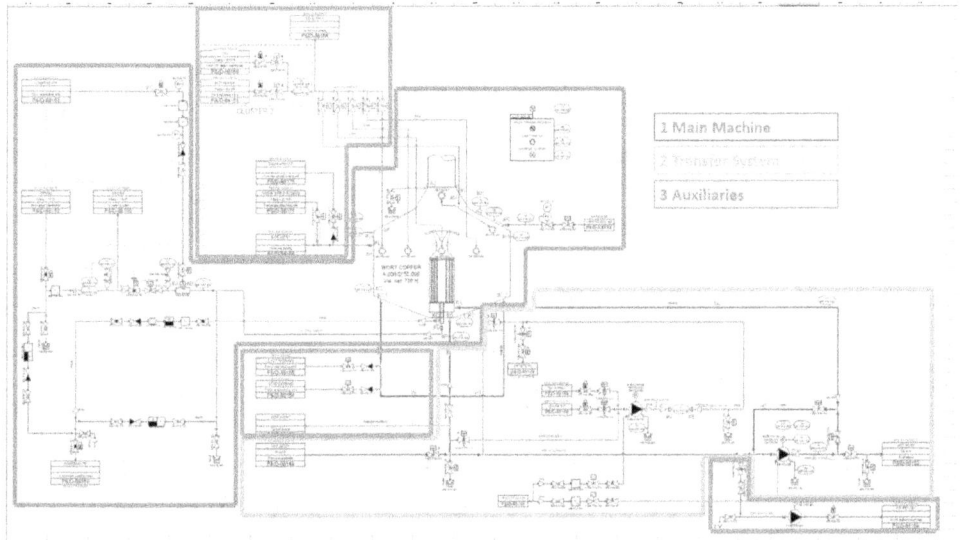

Figure 62: Process and Instrumentation diagram for Wort Kettle showing standardised machine sections, from wort kettle HACS file.

Later when we moved to the Brewhouse, using the general arrangement drawing did not make sense, a better organization for process equipment is to use the Process and Instrumentation Diagram (P&ID), as shown in Figure 62 for the brewhouse wort kettle. The same approach is advocated by Bloom (Bloom, 2006, p. 148) who uses the P&ID to determine system boundaries.

Smith (Smith, 1993, p. 65) pays a lot of attention to defining system boundaries, whilst agreeing with the importance, I found it fairly straightforward to define the boundaries if we are using a P&ID diagram or a general arrangement in the case of packaging machines. Nevertheless, it is a necessary first step and that is why it is the first sheet of the HACS Excel file.

In the brewhouse we stopped using the terms "Infeed" and "Outfeed" as usually the same piping system is used to fill a tank and to empty it, so we just use "Transfer System" instead.

One advantage of using the P&ID is that in theory every component of the asset should appear on the P&ID. However, P&IDs are not normally provided with packaging machines, so there we use the layout drawing.

CONSTRUCTION GROUPS

We developed the concept of Construction Groups to make the execution of maintenance tasks from the Planned Maintenance system more efficient.

We have fairly large and complex machines, so we defined a **CONSTRUCTION GROUP as a set of assemblies or parts in the machine that share a common function.**

The idea was to issue all the Planned Maintenance schedules related to one construction group at one time when possible. This would mean that the lock-out and isolation of that construction group, followed by any disassembly to access the components, would be done once and then all relevant Planned Maintenance schedules completed before that construction group is re-assembled. This is partly driven by the common habit of printing the Planned Maintenance schedules at the start of each month for the coming month.

What we aim to avoid with the use of Construction Groups, is that we could issue a Planned Maintenance schedule to inspect a component, the Technician isolates that section (Construction group), and perhaps he has to dismantle quite a lot of assemblies to get to the component. He does the inspection and reassembles everything. Potentially next week, we may issue another inspection for another component in the same construction group. Then the Technician would be asked to go again to the same section, isolate it and dismantle the same assemblies that he dismantled the previous week, and carry out the next inspection.

The Technician would be right to ask why we did not do the two inspections at the same time, so that he did not have to dismantle the same assembly 2 weeks in a row. And of course, every time we disassemble something we increase the risk of it being incorrectly re-assembled. So, we try to avoid this by working with Construction Groups, issuing all the Planned Maintenance schedules for one construction group at the same time whenever we can, usually on a monthly basis.

Before we introduced this concept, which was first proposed by Mr. Giang in our team, it was required that each and every Planned Maintenance schedule be given an "A", "B" or "C" priority in the Maximo system. Generally, this meant in most breweries that the "A" priority Planned Maintenance schedules were done in the first two weeks of the month, the "B" priority schedules in the 2nd two weeks, and the "C" priority schedules were done if there was sufficient time left over.

It should be clear that the priority approach led to a lot of wasted time and unnecessary dismantling when the Planned Maintenance schedules are in the same Construction Group and that Construction Group is then accessed repeatedly for the different priority schedules.

This is an example of the improvements we can make to the lives of the workers on the shop floor if we use our practical knowledge and experience.

However, the input fields in Maximo require a priority, so now we just put every schedule with the same priority.

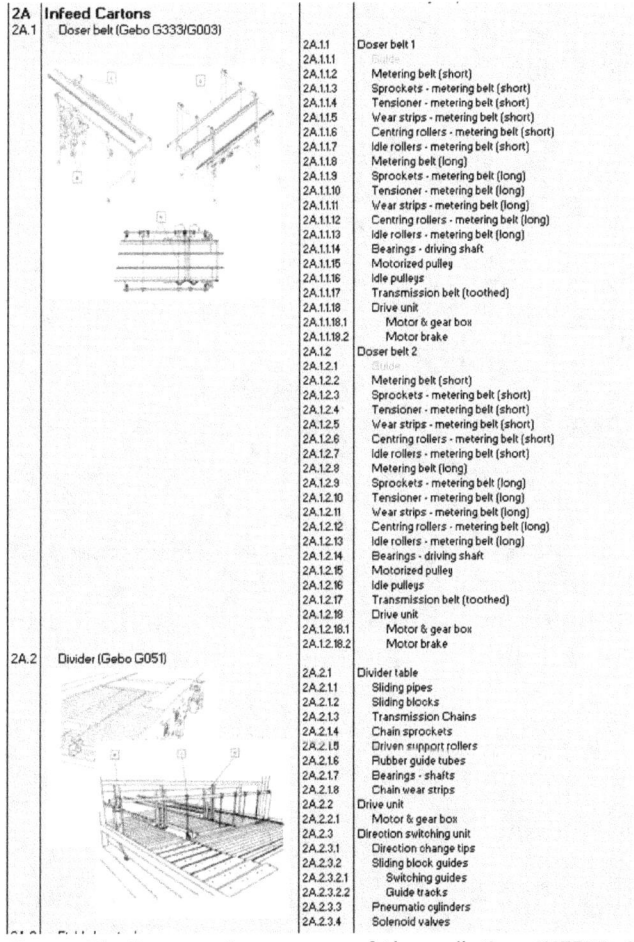

Figure 63: Construction groups of the palletiser INFEED CARTONS section, from palletiser HACS file.

In Figure 63 I show part of the second main worksheet of the Excel file, which we call the HIERARCHICAL STRUCTURE. We can see the construction groups of the palletiser carton infeed (INFEED CARTONS), which is section 2A.

There are two construction groups:

- 2A.1 Doser Belt (Gebo G333/G003)
- 2A.2 Divider (Gebo G051)

If you refer back to Figure 56 in the previous chapter, you can see how each construction group is linked back to the asset number of the palletiser itself.

For each construction group we include a picture, usually from the Spare Parts Catalogue, which helps with visualization.

Next we divide the construction groups into assemblies. An assembly is a group of components that work together.

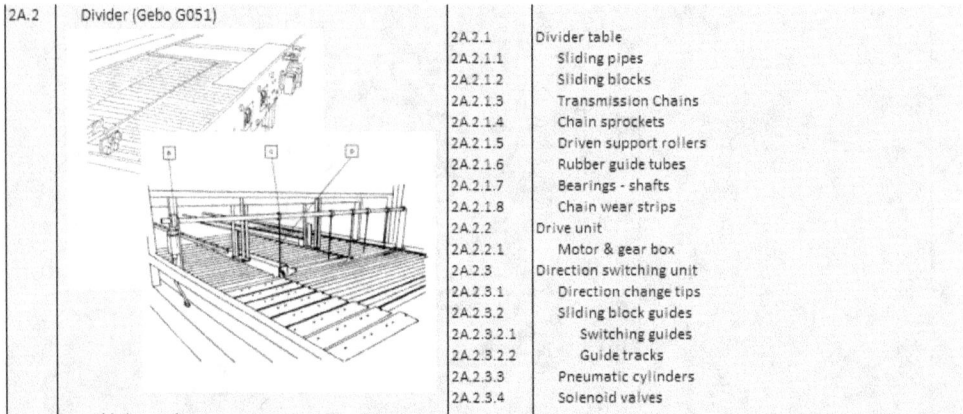

Figure 64: Construction group and assemblies of palletiser hierarchy, from palletiser HACS file.

In Figure 64, the construction group 2A.2 Divider (Gebo G051) has three assemblies:

- 2A.2.1 Divider Table.
- 2A.2.2 Drive Unit.
- 2A.2.3 Direction Switching Unit.

The Item 2A.2.3.2 SLIDING BLOCK GUIDES is a sub assembly with two MSI components.

- 2A.2.3.2.1 Switching Guides.
- 2A.2.3.2.2 Guide tracks.

Note how we indent components of each assembly (Tab to the right) so that it is easier to see which components belong to which assembly or sub assembly.

It is in the listing of the components of the assembly that we actually define the extent of the maintenance system, and it is a critical step. We look at the Spare Parts Catalogue and there we can (hopefully) see all of the components of the assembly (Consider the Spare Parts Catalogue to be an illustrated Bill of Materials). We have to decide which parts to include in the hierarchy as an MSI. **We only include components that are an MSI.**

How do we decide which components are MSIs and which can be left out of the hierarchy?

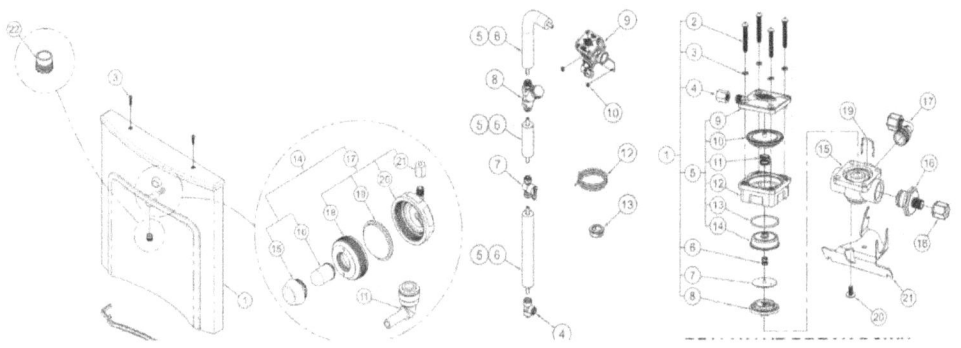

Figure 65: Typical spare parts catalogue with exploded views.

What we do is search through the Spare Parts Catalogue to find the drawing of each assembly, the one shown in Figure 65 is a typical example.

Looking at such a drawing, we can decide which parts are MSIs (or sub-assemblies) to be included in the hierarchy, remembering of course the definition of an MSI from Chapter 8:

A MAINTENANCE SIGNIFICANT ITEM (MSI) is the lowest level component or sub assembly, that is likely to cause a BREAKDOWN of the asset in its useful lifetime, to which we apply a MAINTENANCE STRATEGY.

So, we need to evaluate each part and decide if it is likely to cause a breakdown of the asset in its useful lifetime. In general, we will include bearings, rollers, chains, belts, sprockets, wear-strips, pistons, motors, gearboxes and valves as MSIs (being moving parts subject to wear), and we will include all electronic parts, such as switches, sensors, photocells, control buttons and solenoids.

Each drawing will also have a component list, shown in Figure 66.

ITEM#	PART NUMBER	DESCRIPTION
1	CONTACT FACTORY	HOUSING ASSEMBLY, BF11
2	7014-257-000	BOTTOM TRIM, H202GO
3	0116-123-000	#10-32 x 1" BUTTON HEAD SCREW
4	1895 710 000	1/4" PUSH IN UNION TEE
5	2169-000-000	1/4" O.D. LLDPE TUBING, BLUE
6	7012-055-000	FOAM PIPE INSULATION
7	7000-420-000	1/4 TURN SHUT-OFF VALVE
8	7013-210-001	"Y" STRAINER, BOTTLE FILLER
9	7014-290-005	PNEUMATIC VALVE MOUNTING ASSY
10	0302-011-000	#6-32 UNC HEX NUT
11	1895-709-000	ELBOW, 1/4" PUSH-IN x 1/4" STEM

ITEM#	PART NUMBER	DESCRIPTION
12	2150-003-199	1/8" OD LDPE TUBE, NATURAL, 3' LONG
13	7100-441-000	LOCKING GROMMET,1.188 DIA HOLE
14	2568-160-001	(-PBH) SIDE OUTLET PUSHBUTTON
15	4005 030 199	AIR TROL PUSHBTN ESCUTCHEON S/S
16	4005-031-199	AIR-TROL PUSHBUTTON
17	2566-025-002	SIDE OUTLET PUSHBUTTON SUB-ASSY
18	2566-022-000	ESCUTCHEON RETAINER
19	2566-001-000	AIR DIAPHRAGM
20	2566-055-199	DIAPHRAGM RETAINER SIDE OUTLET
21	1895-450-000	1/8" O.D. NYLON COMPRESSION NUT
22	7013-119-000	NEOPERL LAMINAR NOZZLE

ITEM #	PART NUMBER	DESCRIPTION
1	2563-000-002	DIRECT ACTING ASSEMBLY
2	0116-012-000	#8-32 x 1-1/4" PHILLIPS ROUND HEAD
3	0331-003-000	#8 LOCKWASHERS
4	1895-450-000	1/8" PLASTIC COMPRESSION NUT
5	2563-000-001	DIRECT ACTING MOTOR ASSEMBLY
6	2563-008-000	PILOT ORIFICE PLATE SPRING
7	2563-019-001	PILOT ORIFICE PLATE ASSEMBLY
8	2563-010-001	WATER DIAPHRAM ASSEMBLY
9	2563-007-000	DIRECT ACTING COVER PLATE
10	2563-004-001	DIRECT ACTING DIAPHRAM ASSEMBLY
11	2563-003-000	ACTUATOR SPRING
12	2563-001-000	MOTOR HOUSING
13	0401-026-000	SEPERATOR CUP O-RING
14	2563-002-199	SEPERATOR CUP
15	2563-000-000	VALVE BODY
16	2800-108-001	1/4" INLET ADAPTER ASSEMBLY
17	2570-051-001	1/4" OD PLASTIC ELBOW ASSEMBLY
18	1895-461-000	1/4" OD NYLON NUT
19	0326-100-000	RETAINING CLIP
20	0124-010-000	#10 X 1/2" PHILLIPS TRUSS HEAD
21	7014-1293-199	BF11/12 VALVE MOUNTING BRACKET

Figure 66: Spare Parts Catalogue component list for the drawing in Figure 65.

When we look at the component list we can eliminate parts described as SCREW, SUPPORT, BOLT, WASHER, NUT, LOCKWASHER, SPACER, HOUSING, COVER PLATE etc....that is clearly not a moving part or is unlikely to be subject to failure.

We focus on the parts that are moving and can wear.

For the DIVIDER TABLE ASSEMBLY (2A.2.1) there are only 8 MSIs as shown in Figure 64:

- Sliding Pipes.
- Sliding Blocks.
- Transmission Chains.
- Chain sprockets.
- Driven Support Rollers.
- Rubber Guide tubes.
- Bearings – shafts.
- Chain wear strips.

These parts we will include as MSIs because we believe that they are components that are likely to cause a breakdown of the asset in its useful lifetime.

One of the challenges we face in building the hierarchy is that the spare parts catalogues of the suppliers usually have no hierarchy at all, so searching through to find the assemblies is quite difficult. The feedback that I give to suppliers on how to improve their documentation is always to follow a hierarchical approach in their Spare Parts Catalogue.

FIELD ELECTRICAL

In Figure 63 I showed that the Machine section 2A INFEED CARTONS has 2 construction groups, but actually there is a third that is not shown in that figure (for simplicity): Field Electrical.

The final construction group in each machine section is always Field Electrical, shown in Figure 67.

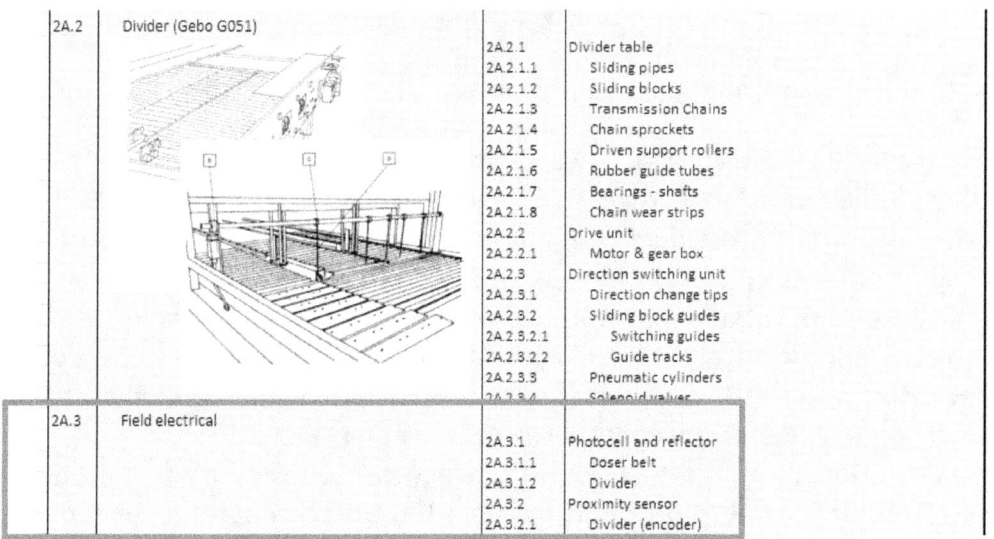

2A.2	Divider (Gebo G051)		
		2A.2.1	Divider table
		2A.2.1.1	Sliding pipes
		2A.2.1.2	Sliding blocks
		2A.2.1.3	Transmission Chains
		2A.2.1.4	Chain sprockets
		2A.2.1.5	Driven support rollers
		2A.2.1.6	Rubber guide tubes
		2A.2.1.7	Bearings - shafts
		2A.2.1.8	Chain wear strips
		2A.2.2	Drive unit
		2A.2.2.1	Motor & gear box
		2A.2.3	Direction switching unit
		2A.2.3.1	Direction change tips
		2A.2.3.2	Sliding block guides
		2A.2.3.2.1	Switching guides
		2A.2.3.2.2	Guide tracks
		2A.2.3.3	Pneumatic cylinders
		2A.2.3.4	Solenoid valves
2A.3	Field electrical		
		2A.3.1	Photocell and reflector
		2A.3.1.1	Doser belt
		2A.3.1.2	Divider
		2A.3.2	Proximity sensor
		2A.3.2.1	Divider (encoder)

Figure 67: Field Electrical components for Machine section 2A Infeed Cartons, from Palletiser HACS file.

We defined Field Electrical components as the electrical devices that are present on the machine, not including any electrical devices in the control panels or MCC (Motor Control Centre). These are devices like photocells and sensors that are installed physically on the machine itself.

However, we decided NOT to include them within the construction groups and assemblies, because they are usually maintained by an electrical specialist, and because they can often be maintained as a group with a single Planned Maintenance schedule. For example, if photocells (Figure 68) are all to be replaced on a Time Based Replacement schedule every 10 years, that can be done as a single Planned Maintenance schedule for the whole set of photocells, as could an inspection of the connections to each photocell.

It is of course possible to argue that the field electrical items should be included in each Construction Group as individual MSIs, but overall, we felt it more efficient to take them out and maintain them as a group, especially as a machine will usually have many identical photocells or sensors on all parts of the machine.

*Figure 68: Photocell, from **author's own collection**.*

In the Field Electrical section, we list the items by component type. First we put Photocell and reflector, then we list each location (in this case Doser Belt and Divider), and then we would continue to list each photocell in this machine section (if there were more).

Next we have Proximity Sensor, and on other machines we would continue with each field electrical component, such as level probes, temperature probes, flowmeters, conductivity sensors, pressure sensors, Emergency stop switches, safety barrier lights, safety door switches, flowmeters etc.

As mentioned before, these are the electrical components found on the machine, and not the components in either the Operator control panel or the Motor Control Centre, which are covered in the machine section 4 of the Hierarchy.

12.5 COMPLETING THE HIERARCHY

The remaining work is to go through the whole machine structure until every section is complete with its construction group and each construction group has all the assemblies and sub-assemblies down to the chosen MSIs that are likely to cause a breakdown of the asset in its useful lifetime.

This process has allowed us to build a hierarchical structure containing only the MSIs of the machine and do note well that we have so far not referred to the function of each MSI or how it may fail.

This approach is a significant deviation from that of Moubray, Smith and Bloom, but it was not done to shorten the RCM process. My way of working, through the hierarchy, is designed to give a more thorough and better structured review of every MSI, which is going to make it very easy when we get to the next step, writing the Planned Maintenance schedules and Job Plans.

I do not believe one could get such a thorough an approach from following a functional approach at a system level.

In building and expanding the HACS over the past 6 years, I have also found that it is very important **not to start with developing the Planned Maintenance schedules until the hierarchy is complete.** This helps maintain the structured approach and prevents wasted work if the hierarchy has to be changed.

I have reviewed personally most of the hierarchies of the machines in the HACS, and it usually takes several attempts from Engineers or students until I am completely satisfied that the hierarchy is following all of the steps described in this Chapter.

The process of building the HACS Hierarchy is shown in Figure 69.

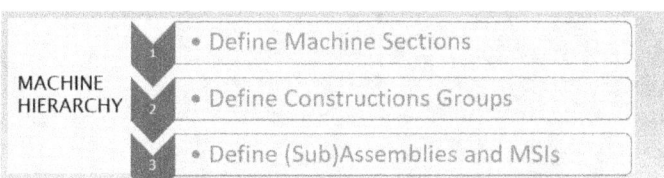

Figure 69: The HACS process so far.

CHAPTER THIRTEEN: WRITING THE HACS PLANNED MAINTENANCE SCHEDULES AND JOB PLANS

Having completed the Hierarchy, we now have a structured list of every MSI for a particular machine and have a reference for each MSI in terms of its assembly, construction group, machine section and the asset (machine) itself.

A MAINTENANCE SIGNIFICANT ITEM (MSI) is the lowest level component or sub assembly, that is likely to cause a BREAKDOWN of the asset in its useful lifetime, to which we apply a MAINTENANCE STRATEGY.

A Planned Maintenance Schedule is a set of Planned Maintenance tasks that are carried out to prevent failures of Maintenance Significant Items. It consists of the Planned Maintenance Schedule that defines the task and frequency, and the Job Plan that describes the maintenance tasks step by step.

So next we need to develop the Planned Maintenance schedules (based on the maintenance strategy) that are needed to prevent a breakdown happening for each MSI, and the logical approach would be to carry out a FMEA for each MSI (see Chapter 9).

FMEA is an effective process to identify every failure mode if you have equipment that you are not familiar with, or the equipment is a completely new design that has not been used before, or if you have a high consequence of failure (Aerospace, petrochemical industries etc..)

But in our case we had already been operating some of the SCALE canning lines for 10 years. Additionally, all of the components of our packaging lines and brewhouses, the bearings, valves, solenoids, motors and pumps come from a limited number of OEM suppliers. Therefore, we should already know many of the failure modes, and we can find out what they are if we just ask the Technicians and Operators.

We do not need to spend hundreds of hours thinking about hypothetical failure modes when we already know the real ones.

Some years ago, I visited the factories of two of the major packaging equipment suppliers in Germany, and I was really surprised that the basic design and construction of the packaging machines was the same as I knew from twenty years earlier when I was a Line Manager in the South African Breweries. There are more safety devices, and there are more electronic control systems, but the basic machine construction has not changed much in most cases.

Even filling valves, a very sophisticated and complex part of the can or bottle filler, are the same basic design that we have used for decades, except that now they have solenoids and probes attached to make them electronic, but most are just iterative improvements of the mechanical valves we used 30 years ago, rather than completely new innovations.

So instead of carrying out a detailed FMEA, we started with the Hierarchy that we had already developed, and then held a review of each MSI in sequence with a cross functional team, making sure that we had in the room Technicians and Operators who knew the machine well.

We looked at each MSI in turn, following the hierarchy of the machine and asked how it can fail or has failed in the past. This process was always facilitated by the working group leader. At no point did we insist on each column of a FMEA table being completed, as this is unnecessary when the function and likely functional failure is clear, and we can go straight to deciding the appropriate maintenance task to prevent the failure.

Then we designed a **Planned Maintenance** schedule for that potential failure, using the Maintenance Task Decision Flowchart, to establish the correct maintenance strategy.

We decided that the minimum frequency of a Planned Maintenance schedule will be monthly. Any activity more frequent than that, such as weekly, is to be included as a CILT (Cleaning, Inspecting, Lubricating, Tightening) activity carried out by operators in the weekly CILT task and not from the Planned Maintenance system.

Moubray (Moubray, 1991, p. 286) and Regan (Regan, 2012, p. 225) rightly caution against taking a light approach to RCM, and I do not believe that is what we did. What we did is spent a lot less time on FMEA tables and a lot more time on developing good Planned Maintenance schedules and Job Plans, as we can use these to do maintenance tasks.

Had we produced all of the FMEA tables recommended I don't know what we would do with them except file them. But I do fully agree that for any critical or hazardous application, then the FMEA does need to be carried out exhaustively and thoroughly documented to ensure safety.

Regan proposes an "RCM solution" of applying the full RCM process to some assets (Regan, 2012, p. 225), and to other less critical assets to apply either Reverse Engineering (where existing manufacturer documents and existing maintenance schedules are reverse engineered to identify failure modes) or just apply lubrication, cleaning and servicing to other assets.

This is a compromise worthy of evaluation, but none of the author's I have examined (except Nowlan and Heap) have, in my view, sufficient focus on the writing of the Preventive Maintenance schedule and Job Plans, instead they over emphasize the identification of failure modes. (Refer to chapter 6 and the inspirational lecture of Bill Hughes that I attended many years ago).

13.1 PALLET CHAIN CONVEYOR: BUILDING THE PLANNED MAINTENANCE SCHEDULE

As an example, I will explain in detail the development of the Planned Maintenance schedules and Job Plan for the pallet chain conveyor. This is a device that uses chains to move either empty or full pallets through the machine and is found in a few places in the palletiser, shown in Figure 70.

Figure 70: Pallet chain conveyor, from author's own collection.

If you look at the hierarchical structure of the pallet chain conveyor (3.1.1) in Figure 71, you see that it consists of only 9 MSIs.

- Chains.
- Chain Wear strips.

- Sprockets.
- Tensioners.
- Bearings.
- Motor & Gear Box.
- Linear bearing sleeves.
- Pneumatic cylinder.
- Solenoid valve.

The last 3 MSIs are part of the stopper mechanism which is marked as "optional" because it is only installed on some of the conveyor sections, usually those at the end of the conveyor section.

Note again for the sake of absolute clarity that we do not include nuts and bolts, support feet, brackets or even the shafts connecting the sprockets. (We discussed the shafts and agreed that they are nearly indestructible, and we have never seen one break, so they are not included as an MSI).

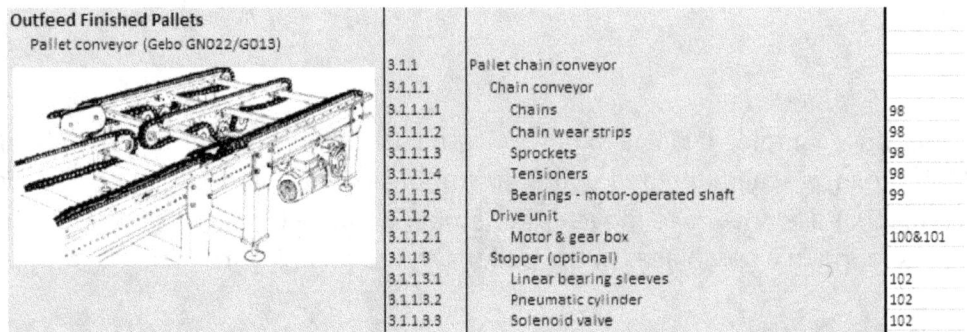

Outfeed Finished Pallets		
Pallet conveyor (Gebo GN022/GO13)		

3.1.1	Pallet chain conveyor	
3.1.1.1	Chain conveyor	
3.1.1.1.1	Chains	98
3.1.1.1.2	Chain wear strips	98
3.1.1.1.3	Sprockets	98
3.1.1.1.4	Tensioners	98
3.1.1.1.5	Bearings - motor-operated shaft	99
3.1.1.2	Drive unit	
3.1.1.2.1	Motor & gear box	100&101
3.1.1.3	Stopper (optional)	
3.1.1.3.1	Linear bearing sleeves	102
3.1.1.3.2	Pneumatic cylinder	102
3.1.1.3.3	Solenoid valve	102

Figure 71: Hierarchical structure of pallet chain conveyor, from palletiser HACS file.

Starting with the first MSI: 3.1.1.1.1. the CHAINS, we know that the most likely failure mode of a chain is that they are subject to stretch, if they stretch too much they can jump off of the sprocket (Most what we refer to as "stretch" is actually wear on each of the individual pins and bushes of the chain).

Now we refer to the Maintenance Task Decision Flowchart, (From Chapter 8, repeated in Figure 72), and ask if it is possible to inspect for pending failure. Happily, it is, there are chain gauges available that make this a very easy task, such as the FB Ketten chain gauge (see Chapter 15, Tools and Calibration). Following the flowchart, as the gauge is quick and easy to use, we decide that the inspection cost is lower than the replacement cost, and as we don't know an easy or economic way to monitor condition on a chain, we decide on a periodic Inspection for the chains.

Next we need to estimate the frequency of the Inspection. It is actually dependent on the length of the pending failure (P-F) maintenance window of the failure curve (Figure 26 in Chapter 6): We have to carry out the inspection when the pending failure can be measured but before the failure occurs.

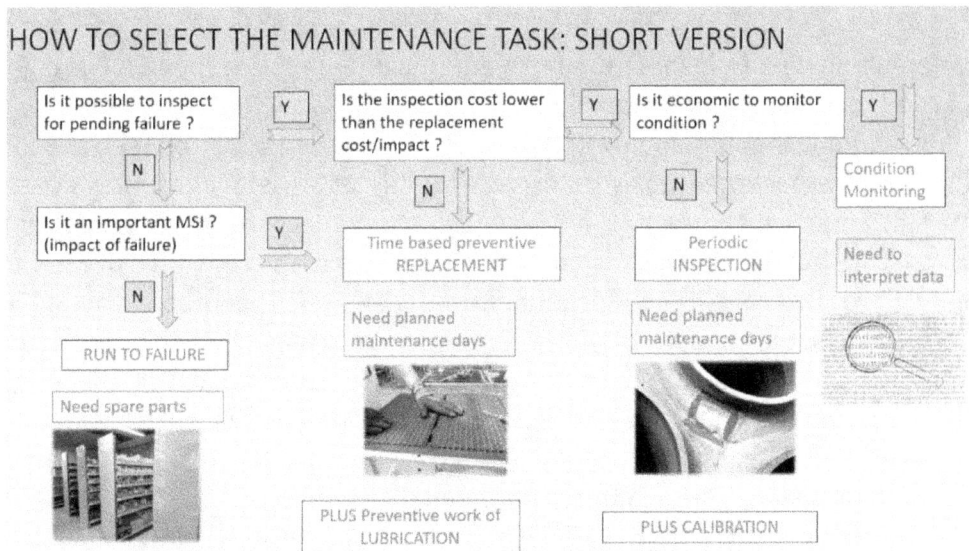

Figure 72: Maintenance Task Decision Flowchart, from Author's own files.

These chains suffer a lot of stress, so we decided to inspect them every 2 months, as they are often replaced after only one or two years of use.

So, we are going to create a Planned Maintenance schedule that is an inspection task to be carried out on the chain every 2 months.

Next we need an inspection standard: The standard for the inspection is easy to define in this case, as for chain stretch the generally accepted standard is to replace chains when the stretch is 3% of the original. The FB Ketten gauge actually measures the stretch for the Technician, so we avoid the possibility of errors that would be likely to occur if we asked the Technician to measure the stretch in millimeters and calculate the percentage stretch.

Before we complete the Planned Maintenance schedule and write the Job Plan, it is always a good practice to look at the next MSI and see if other maintenance tasks should also be combined in the Job Plan, for purposes of packaging the work. We only do this if the next MSI has the same Maintenance Strategy and frequency, otherwise we make a new Planned Maintenance schedule. (We do not combine Planned Maintenance schedules that have different frequencies or different maintenance strategies, as when we do that we end up carrying out an overhaul, which is not effective Planned Maintenance, see Chapter 3).

The next MSI is 3.1.1.1.2 the CHAIN WEAR STRIP. Again, we follow the Maintenance task decision flowchart (Figure 72) and reach the conclusion that the wear-strip is also easy to inspect, especially if we do it at the same time as we inspect the chain, because you just measure the thickness of the remaining wear-strip.

In this case we think that the life of the chain and wear-strip are not too different, so we can inspect them together (same frequency) and we are going to combine them into one job plan.

But what is the standard for the wear on the wear-strip? Often I see a Job Plan that says *"inspect the wear-strip for wear"* without any standard. How does the Technician judge if it is OK or not? We can easily measure the thickness of the wear-strip, but if we want to specify the minimum in mm it will be different for nearly every wear-strip and location. So, to make this repeatable, the Steering Committee set a standard that all wear-strips are replaced when they are worn down to 50% of the original thickness. (If you want to check the original thickness, you can measure a new piece in the spare parts store).

Using this simple standard considerably streamlines the information and decision making required in the maintenance system.

The next MSI is 3.1.1.1.3 the SPROCKET. As it is integrated with the chain, we are going to include it in the same Planned Maintenance schedule as well. Referring to the Maintenance Task Decision Flowchart, the first question is "Is it possible to inspect for pending failure", and the answer is yes, it is possible, but it is very difficult to describe the standard.

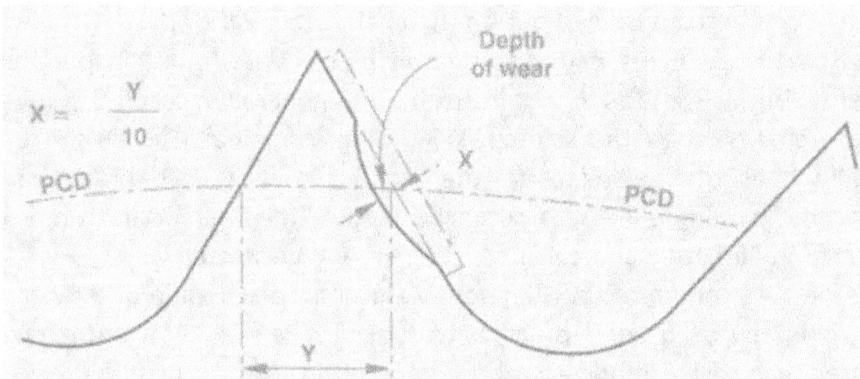

Figure 73: Measuring wear on a sprocket tooth, image from www.Renold .com

As shown in Figure 73, the wear on a sprocket is on the PCD (Pitch Circle Diameter) of the chain rollers where the stress applied by the rollers in the chain wears away at the sprocket face. The depth of wear (X) should be less than 10% of the original width of the sprocket tooth at that point.

There are sets of gauges available to measure sprocket wear, but they are quite difficult to use, so we settled on a logical work around for this inspection: We don't inspect sprockets!! What we do is we inspect the chain, and when the chain is stretched beyond limits we replace the chain AND the sprocket. This is completely correct as you should never put a new chain on an old sprocket, so the need to inspect sprockets is avoided.

The next MSI in Figure 84 is 3.1.1.1.4 the TENSIONER. This is a screw that moves the entire motor and drive shaft backwards or forwards to create more or less tension on the chain and take up any slack that is in the chain.

Here the failure mode would be that the chain is not tensioned, and as it is also an integral part of the chain and sprockets assembly we decided to include this in the same Inspection Planned Maintenance schedule as well.

Referring to the Maintenance Task Decision Flowchart, it is quite clear that for chain tension we would need to carry out an Inspection task.

We are going to need a standard for chain tension, and that is normally defined as how much you can move the chain across a line perpendicular to its path of travel. Frequently this is done by hand, a technician will usually try to move the chain and decide if it is sufficiently tensioned. There are tools available that apply a force to the chain and the movement amount for a particular force can be measured. We felt that this is unnecessary for most chains, but we still need some sort of standard.

So, we defined the standard as the amount of movement when the chain is moved by one hand of the Technician, (knowing that this will vary with the different strengths of different Technicians). The total mid-span movement in the slack span of the chain should be approximately 5% of the driving and driven sprockets center distance.

Do note that chain tension is different to chain stretch (elongation).

So far we have decided that we will incorporate into one Planned Maintenance schedule an inspection of the chain stretch, chain tension, wear-strip and tensioner every 2 months, so we write the Planned Maintenance schedule as per Figure 74.

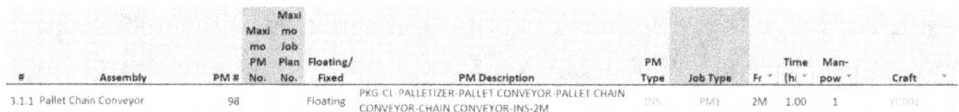

#	Assembly	PM #	Maxi mo PM No.	Maxi mo Job Plan No.	Floating/ Fixed	PM Description	PM Type	Job Type	Fr	Time (hr)	Man-pow	Craft
3.1.1	Pallet Chain Conveyor	98			Floating	PKG-C1-PALLETIZER-PALLET CONVEYOR-PALLET CHAIN CONVEYOR-CHAIN CONVEYOR-INS-2M	INS	PM1	2M	1.00	1	ELPO2

Figure 74: PM 98 for pallet chain conveyor, from palletiser HACS file.

The Planned Maintenance schedule is the description of the maintenance activity with its frequency and the type of maintenance strategy. It is the WHAT and the WHEN of the maintenance strategy. It is now entered in the PM & Job Plans worksheet of the HACS Excel file.

(The HACS Excel File has a first worksheet that is the structural overview showing the machine sections, a second worksheet that is the hierarchy of the machine, described in the previous chapter, and the third worksheet is the PM & Job Plans worksheet, which has the same format of columns as the upload sheets, so that once the Planned Maintenance schedules and Job plans are written it is easy to transfer them all into the upload sheets required by Maximo).

In the first column, the reference 3.1.1 links the **Planned Maintenance schedule** to the chain conveyor. The Planned Maintenance schedule number is just used for organization within the file.

Fixed or floating refers to whether the Planned Maintenance schedule is applied whenever a certain period is reached (floating) or is a set calendar time such as the first day of the month (fixed).

The Planned Maintenance schedule name is standardised as *Department-area-machine-construction group-assembly-PM type-period*.

Job type is selected from the following options:

PM1   INSPECTION.
PM2   LUBRICATION.
PM3   CALIBRATION.
PM4   REPLACEMENT.
PM6   CONDITION MONITORING.
PM7   CLEANING.
PM8   SAFETY INSPECTION.

We fill in the frequency in terms of months or years. Currently our Planned Maintenance schedules in Maximo are all defined as a calendar frequency (weekly, monthly etc..). We assume that the production operation is 5 days a week and 24 hours per day in the HACS when we assign the frequency of the Planned Maintenance schedule. However, this means that when an asset is utilized more or less frequently than the envisaged 5 days and 24 hours, for example if we go to 5 days and 12 hours, we then have to manually adjust the frequency of every Planned Maintenance schedule. We are developing the ability to link the Planned Maintenance schedules to the operating hours of the machine, but this is an extensive change that will take some time to complete.

In the "time" and "manpower" columns we estimate the number of people needed to carry out the task and the time that they will need. This allows us to schedule the work and to calculate the resources needed. Normally we cannot do this until we have written the Job Plan.

For each HACS file we total the manpower requirements and include this as a separate worksheet (fourth worksheet in the HACS Excel file) so that we can plan throughout the year the resources needed for the execution of the Planned Maintenance schedules. This is explained in more detail in Chapter 19.

In the final "Craft" column we assign the Planned Maintenance schedule to a profession from this LIST:

YC007	CAR MECHANIC.
YC008	QUALITY TECHNICIAN.
YC009	MAINTENANCE ENGINEER.
YC010	WELDER.
YC011	OPERATOR.
YC012	LUBRICATOR.

WRITING THE JOB PLAN

Having completed the Planned Maintenance schedule, we now need to write the Job Plan. The Job Plan is a sequential list of instructions that is printed with the Planned Maintenance schedule for the Technician to follow in order to carry out the work (Recently we have started to use a mobile version of Maximo that allows the Technician to see the Job Plan on a mobile device).

Whereas the Planned Maintenance schedule was the WHAT and the WHEN, the Job Plan is the HOW of the Maintenance Strategy.

We normally number the steps of the Job Plan in increments of 10 to allow us to add any missed steps later without having to renumber all of the intervening steps.

As mentioned earlier, Maximo is only able to handle text descriptions in the Job Plan, attaching pictures or figures is not supported.

Sometimes the Job Plan is simple, and a few lines will describe the task adequately. But often we need to give a more detailed instruction as to what the Technician needs to do, and ideally to show some pictures or diagrams. We have to include these as a separate SOP. Our SOPs are developed and shared in Swipeguide, (Swipeguide, 2013) which is explained in the next chapter.

A key strategic decision when building a Planned Maintenance system is to what level of detail will you write the Job Plans? Will you write them so that they can be understood and carried out by anyone, regardless of their knowledge or experience? In this case you will need to describe every small step, such as how to use tools and gauges. Or you could write them assuming that the Technician already knows the machine and has worked on it for some time, but in this case the Technician with that level of competence has to always be available.

The Steering Committee decided that we should write all of our Job Plans assuming that we had a trained and qualified Technician (so he/she is able to use tools, take measurements etc.) but that he/she was not familiar with the specific machine. Therefore, everything related to the specific machine's design and construction is explained assuming the Technician is not familiar with the machine, but generic maintenance and measurement activities are not described, such as how to remove a sprocket from a shaft or how to measure the thickness of a wear strip.

Defining this was an important decision that has avoided the need for specific training of Technicians on each machine as that knowledge is built into the Job Plans.

Figure 75 is the Job Plan (Job task description) for the Planned Maintenance schedule to inspect the chain stretch, chain tension, wear-strip and tensioner every 2 months. Note well that in every Job Plan we include the reference to any SOP and a reference to the spare parts needed for that task.

Job Task #	Job Task Description	SOP/ Attachments	Spare Part Name	Quantity	Part Number
10	LOTO machine				
20	Check And Adjust Chain Tension As Per Standard SOP: INS-4	INS-4	Chain (Short)	3	04299970205
30	Check Elongation Of The Chain As Per Standard SOP: INS-3	INS-3	Tensioners	2	TRP152
31	Replace Chain, Tensioners And Sprockets If Elongation Is More Than 3%	REP-12 Rep-14	Sprocket - Chain Ends	6	04299960002
32	As Per Standard SOP: REP-14, 12 And GC-CL-PL-086	GC-CL-PL-086	Sprocket - Motor-Operated Shaft	3	04304762401
40	Inspect Chain Wear Strips For Wear And Tear As Per Standard SOP: INS-13	INS-13	Chain Wear Strips (Short)	3	04299961201
41	Replace If Necessary As Per SOP: GC-CL-PL-084	GC-CL-PL-084			
50	Clean And Lubricate All Chains As Per Standard SOP: LUB-2	LUB-2			
60	Operate Pallet Chain Conveyor In Auto Mode				
70	LOTO Release, Test The Machine And Hand-over To Operator				

Figure 75: Job Plan for Palletiser pallet chain conveyor PM 98, from Palletiser HACS file.

Every job plan should start with a LOTO instruction. LOTO, (Lock-Out, Tag-Out), is the process of locking the energy sources of a machine (electrical, mechanical, air etc.) and identifying the locks with TAGS (identification sign) so that no-one can inadvertently activate an energy source in the machine and injure the Technician or Operator.

Another strategic decision that the Steering Committee had to make was whether to describe the LOTO steps for each machine or construction group before starting the Job Plan. One could argue that it is necessary to be very specific about the LOTO requirements to avoid any possibility of an accident. However, each machine has a LOTO procedure developed specifically for that machine, and it is displayed as a LOTO standard at the machine itself. Including that procedure in the Job Plan would be repetitive and would make it necessary to update the Job Plan whenever the LOTO standard is revised.

So, we just state "LOTO machine", knowing that the LOTO standard is at the machine. In respect to the earlier assumption that the Technician is a qualified Technician unfamiliar with the machine, we assume that he/she is able to follow the LOTO standard on the machine. **If you are developing your own Planned Maintenance system, you need to make sure that the necessary LOTO standards are in place.**

In line 20 of the Job Plan we give the instruction to check the chain tension. Here we just refer to an SOP: INS-4, which is a generic SOP. (We developed generic SOP's because we knew that there would be identical activities that are carried out in many Job Plans. The numbering format of the SOPs is described in sections 14.1 and 14.2). Likewise for the next item, to check the chain elongation (stretch), we also refer to the generic SOP: INS-3.

In the event that the chain is stretched we will need to replace it, so we refer to the SOPs for that in the next line, REP-12 and REP-14 and GC-CL-PL-086.

In line 40 we instruct the Technician to inspect the wear strips, referring to SOP: INS-13, and how to replace them if necessary using SOP: GC-CI-PL-084.

Whenever there is an inspection, we always must include the instructions necessary to carry out the replacement in the case that the inspection reveals that to be necessary.

In line 50 we have added a lubrication task, to add oil to the chain and sprockets. This needs to be done every 2 months so instead of making a separate Planned Maintenance schedule to lubricate the chain, we included it in the inspection schedule. LUBRICATION appears at the bottom of the Maintenance Task Decision Flowchart and has to be remembered to be included for each MSI where needed, as does Calibration. (I did state earlier that we do not combine Planned Maintenance schedules that have different frequencies or Maintenance strategy, but in the case of LUBRICATION it is OK to add it to an inspection).

In line 70 we remove the LOTO, and the Job Plan is completed.

This is the first Planned Maintenance schedule and Job Plan of this section, for the chains, wear strips, sprockets and tensioners.

Now we go to the next MSI in Figure 84, 3.1.1.1.5 the BEARINGS on the shafts. If we refer to the Maintenance Task Decision Flowchart, the first question is whether it is possible to inspect for pending failure. The answer is YES, it is possible.

But to inspect a bearing is not so easy. Some people like to specify that you can feel the amount of play on the shaft, but at that point the bearing is already well out of the Pending Failure (P-F) part of the failure curve and firmly in the performance loss part of the curve. You can use ultrasound to listen to the bearing, but pallet conveyor shaft bearings move at a very slow speed so ultrasound will not give any useful data. You can dismantle it and look carefully at the ball races, but then we have moved to the point where the cost of inspection is no longer lower than the replacement cost.

So, at the "cost of inspection" question, for most bearings we move down to TIME BASED REPLACEMENT, and we now need to estimate the lifetime. (We may still have inspection for very large/expensive bearings, or more likely Condition Monitoring).

In this case we have an estimate of 5 years for the life of the bearing, so we make a Planned Maintenance schedule for a TIME-BASED REPLACEMENT of all the bearings every 4 years.

#	Assembly	PM #	Maxi mo PM No.	Maxi mo Plan Job No.	Floating/ Fixed	PM Description	PM Type	Job Type	Fr	Time (hr	Man- pow	Craft
3.1.1	Pallet Chain Conveyor	98			Floating	PKG-CL-PALLETIZER-PALLET CONVEYOR-PALLET CHAIN CONVEYOR-CHAIN CONVEYOR-INS-2M	INS	PM1	2M	1.00	1	YC001
3.1.1	Pallet Chain Conveyor	99			Floating	PKG-CL-PALLETIZER-PALLET CONVEYOR-PALLET CHAIN CONVEYOR-CHAIN CONVEYOR-REP-4Y	REP	PM4	4Y	3.00	2	YC001
3.1.1	Pallet Chain Conveyor	100			Floating	PKG-CL-PALLETIZER-PALLET CONVEYOR-PALLET CHAIN CONVEYOR-DRIVE UNIT-REP-4Y	REP	PM4	4Y	1.00	1	YC001
3.1.1	Pallet Chain Conveyor	101			Floating	PKG-CL-PALLETIZER-PALLET CONVEYOR-PALLET CHAIN CONVEYOR-DRIVE UNIT-LUB-2Y	LUB	PM2	2Y	2.00	1	YC001
3.1.1	Pallet Chain Conveyor	102			Floating	PKG-CL-PALLETIZER-PALLET CONVEYOR-PALLET CHAIN CONVEYOR-STOPPER-REP-5Y	REP	PM4	5Y	1.00	1	YC001

Figure 76: Planned Maintenance schedule 99 for pallet chain conveyor from palletiser HACS file.

The Planned Maintenance schedule for the Time-Based replacement of the bearings is highlighted in Figure 76, and the Job Plan for this replacement is shown in Figure 77. This is a very short Job Plan, only containing the instruction to replace the bearings as per the SOP: REP-7.

Job Task #	Job Task Description	SOP/ Attachments	Spare Part Name	Quanti ty	Part Number
10	LOTO Machine	REP-7	Bearings - Motor-Operated Shaft	3	00000184816
20	Replace Bearings Of The Motor-Operated Shaft As Per Standard SOP: REP-7				
30	LOTO Release, Test The Machine And Hand-over To Operator				

*Figure 77: Job Plan for **Planned Maintenance** schedule 99.*

Referring back to the hierarchy of the pallet chain conveyor (Figure 71), we have completed the Planned Maintenance schedules and Job Plans for the assembly 3.1.1.1 Chain Conveyor, and the next assembly is the drive unit: 3.1.1.2. which consists only of the single MSI: Motor & Gearbox.

It is important to have the right maintenance strategy here, as we have several hundreds of conveyor Motor & Gearbox units in a typical packaging hall.

We decided to treat the motor and gearbox as a single MSI. Let's consider how the motor & gearbox can fail:

Firstly, it will fail if there is no oil in the gearbox (Failure mode: No oil). One of the common Planned Maintenance schedules we see in our breweries is an inspection to check the gearbox oil.

As already described in Chapter 8, the oil in the gearbox cannot disappear, if the gearbox leaks the oil has to run onto the floor. So, in the HACS we rely on the operators to report and raise a TAG if there is oil on the floor (and we include this in the weekly CILT check), rather than wasting resources checking every gearbox oil level.

To prevent premature failure of bearings or gears (Failure mode, dirty oil) we do need to change the oil to make sure that we remove any accumulated metal particles and when we do that we should check that there is no water mixed into the oil. So, we have a Planned Maintenance schedule to drain and replace the oil every 2 years (This could be a higher frequency for other gearboxes depending on the stress level). At this point you can consider if you want to carry out oil analysis, a basic form of Condition Monitoring. In this case you would send an oil sample to a laboratory, and they will provide a report on the metal content and other undesirable matter in the oil, such as water or acids. We do not specify this yet as the cost of the analysis is quite high relative to these small gearboxes, but it is certainly useful for larger gearboxes.

One further failure mode for the Motor & Gearbox is that the Motor "burns out". If we return to the Maintenance Task Decision Flowchart, we can again ask, "Is it possible to inspect for pending failure?". The insulation in the motor windings breaks down over time causing low resistance and subsequent overheating and failure. We cannot really inspect for this, as we have no way of knowing the length of the Pending Failure (P-F) section of the failure curve, but it seems unlikely that we would be able to detect the lower resistance before the motor burns out.

Another failure mode would be failure of the internal bearings in the motor. We cannot inspect for this without dismantling the motor, unless we apply Condition Monitoring, for our larger brewhouse motors we do use vibration sensors to monitor the bearings.

We came to the conclusion that for small conveyor motors (these are about 0.75kW) it is not economic to inspect for electrical "burn out" or bearing failure, and for the smaller motors we will not apply Condition Monitoring, so we end up with TIME-BASED REPLACEMENT of the motor and gearbox every 4 years (PM 100) and changing the oil in the gearbox every 2 years (PM 101), as shown in Figure 79.

#	Assembly	PM #	Maximo PM Plan No.	Maximo Job Plan No.	Floating/Fixed	PM Description	PM Type	Job Type	Fr	Time (hi	Man-pow	Craft
3.1.1	Pallet Chain Conveyor	98			Floating	PKG-CL-PALLETIZER-PALLET CONVEYOR-PALLET CHAIN CONVEYOR-CHAIN CONVEYOR-INS-2M	INS	PM1	2M	1.00	1	YC001
3.1.1	Pallet Chain Conveyor	99			Floating	PKG-CL-PALLETIZER-PALLET CONVEYOR-PALLET CHAIN CONVEYOR-CHAIN CONVEYOR-REP-4Y	REP	PM4	4Y	3.00	2	YC001
3.1.1	Pallet Chain Conveyor	100			Floating	PKG-CL-PALLETIZER-PALLET CONVEYOR-PALLET CHAIN CONVEYOR-DRIVE UNIT-REP-4Y	REP	PM4	4Y	1.00	1	YC001
3.1.1	Pallet Chain Conveyor	101			Floating	PKG-CL-PALLETIZER-PALLET CONVEYOR-PALLET CHAIN CONVEYOR-DRIVE UNIT-LUB-2Y	LUB	PM2	2Y	2.00	1	YC001
3.1.1	Pallet Chain Conveyor	102			Floating	PKG-CL-PALLETIZER-PALLET CONVEYOR-PALLET CHAIN CONVEYOR-STOPPER-REP-5Y	REP	PM4	5Y	1.00	1	YC001

Figure 78: Planned Maintenance schedules 100 and 101 for pallet chain conveyor from palletiser HACs file.

For these two Planned Maintenance schedules the job plans are shown in Figure 79.

Another strategic decision we had to take was how much of the total maintenance activity we need to describe. For the 4-year Time Based Replacement of the Motor & Gearbox, we specify that we replace the unit with a spare motor from the (Spare Parts) store, and then we overhaul the unit that was removed.

Job Task #	Job Task Description	SOP/Attachments	Spare Part Name	Quantity	Part Number
10	LOTO Machine	GC-CL-PL-017	Gear Motor	1	00000241185
20	Replace Gear Motor With Spare Motor From Store As Per SOP: GC-CL-PL-017				
30	Overhaul The Replaced Motor (At Workshop)				
40	LOTO Release, Test The Machine And Hand-over To Operator				
10	LOTO Machine	LUB-1			
20	Change Oil In Gearbox As Per Standard SOP: LUB-1				
30	LOTO Release, Test The Machine And Hand-over To Operator				

Figure 79, Job Plans for PM's 100 and 101 for Pallet Chain Conveyor form palletiser HACS file.

As mentioned in section 8.1 we decided not to describe maintenance tasks carried out in the workshop or off-line, and only to describe those that are carried out on the line itself. Partly this is for expediency, and partly because this sort of work is often carried out by third parties. In many locations if a motor is to be overhauled it is sent out to a specialised workshop that will rewind the motor, fit new bearings and return it as an overhauled unit.

So, for tasks that are off-line and likely to be carried out by a third party we do not describe them in detail. However, some tasks that are off-line but normally carried out in house, such as a filler valve overhaul, we do describe fully in the Job Plan or SOP.

Having completed the Planned Maintenance schedules and job plans for the Drive Unit, 3.1.1.2, the final assembly is the STOPPER, 3.1.1.3. the Planned Maintenance schedule and Job Plan is developed for this unit in the same way as the ones already described.

13.2 AM AND PM INTEGRATION

In the TPM system, we aim to improve our organizational capability and to transfer maintenance tasks from Technicians in the Engineering Team to trained Operators which is then referred to as Autonomous Maintenance (AM).

As breweries start the TPM process, in the journey to TPM Bronze they have to carry out steps (APM steps 1-3) that include documenting standards, removing sources of dirt, standardising the cleaning and beginning to take over tasks such as lubrication.

After TPM Bronze is achieved breweries move to the advanced level and the first step (APM step 4) is transferring maintenance tasks from the Engineering Technician to the Operator, in fact the roles of Technician and Operator become combined into one.

In building the HACS we realized that our many breweries all have different levels of TPM maturity, so we were not able to say which Planned Maintenance schedules were to be carried out by an Engineering Technician and which by an Operator. Therefore, in the PM's we assign every Planned Maintenance task to an Engineering Technician (YC009 Maintenance Engineer in Maximo codes) and allow each brewery to decide when that Planned Maintenance schedule execution can be transferred to the operator (YC011) once the required training and handover of the task has been completed.

13.3 VERIFICATION AND VALIDATION

Sometimes during development of the Planned Maintenance schedules, we were not completely sure about the specific steps in a Job Plan, or about the frequency of a maintenance strategy. For example, if we were not sure about the normal lifetime of a bearing or how to inspect a particular part for pending failure.

We would need to ask the supplier for more information, but we did not want to contact the supplier each time we had a question in the Planned Maintenance schedule development process.

So, to make sure that we kept moving forwards we adopted the practice of highlighting each issue in the Planned Maintenance schedule or Job Plan where we had a question, and then moving on with the next item.

When we had completed developing the Planned Maintenance schedules for a machine, or more often for a complete packaging line, we then set up a verification exercise with the supplier. This involved having an expert from the supplier spend some days at the brewery (at our expense) where we would ask them to advise us on each of the points that we had previously highlighted.

This allowed us to verify the Planned Maintenance schedules and added a lot of value to the development of the HACS.

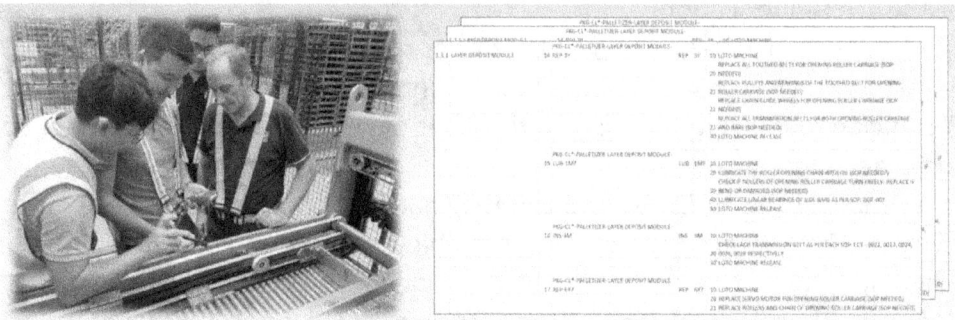

Figure 80: Verification of Palletiser with Gebo expert.

Some of the verification was done at the machine, in Figure 80 Mr. Gianni from Sidel/Gebo is advising the HACS development team on the way to measure the tension of a toothed belt.

Regan describes a more structured validation process in Chapter 11 of her book (Regan, 2012, p. 185), which is a process to review the Working Group's decisions and acts as a Quality Control process.

Regan describes a validation team and the use of a structured validation package and validation checklist to check the FMEA, functions, failure modes, maintenance tasks and strategies produced by the working group.

As the HACS is rolling out globally, individual breweries are using the database files for the SCALE machines to develop HACS for their own similar machines locally.

Some of these files are not written to the same standard as the original HACS, there is sometimes a tendency to take shortcuts. There is no Steering Committee controlling the local HACs development at the individual breweries, an area that we are working to improve.

Currently the Global Maintenance team is setting up a structured validation process, similar to that described by Regan, where a Validation Committee will meet and review HACS files submitted by local breweries against a validation checklist before they are uploaded to the HACS database. Global Maintenance are providing great assistance in validating the local HACS files and certifying Engineers who have completed the HACS training modules which are described in Chapter 21.

13.4 SUMMARY OF PLANNED MAINTENANCE SCHEDULE DEVELOPMENT

In Chapter 12 we described the steps of developing the HACS Hierarchy.

Now we can complete the picture of the HACS development process by adding the 2nd stage, the development of the Planned Maintenance strategy and Job Plan per MSI, as shown in Figure 81.

This Figure of the HACS development process replaces the TPM Maintenance Standards Team Route in Chapter 5, Figure 19. It is a little narrow, in that it assumes no existing HACS is available to adapt, and it focuses only on the actual HACS schedule and not on the other maintenance requirements like spare parts availability or Planned Maintenance days. However, the flowchart does show clearly the steps in developing the HACS Planned Maintenance schedules, separating the Hierarchy from the Planned Maintenance strategy, and defining the outputs of the process.

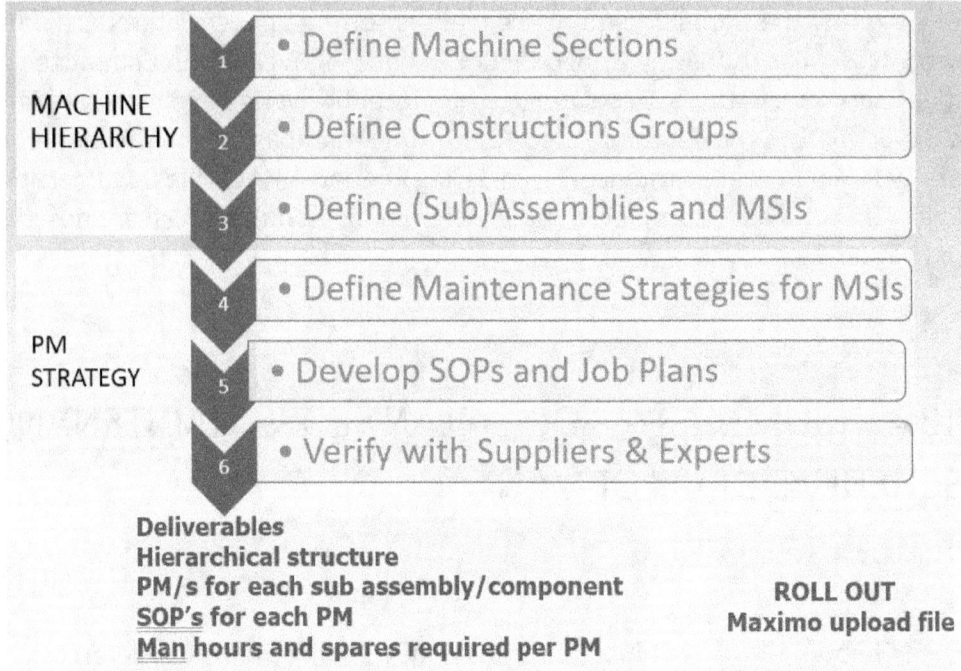

MACHINE HIERARCHY

1 • Define Machine Sections

2 • Define Constructions Groups

3 • Define (Sub)Assemblies and MSIs

PM STRATEGY

4 • Define Maintenance Strategies for MSIs

5 • Develop SOPs and Job Plans

6 • Verify with Suppliers & Experts

Deliverables
Hierarchical structure
PM/s for each sub assembly/component
SOP's for each PM
Man hours and spares required per PM

ROLL OUT
Maximo upload file

Figure 81: HACS development process from author's HACS training files.

Our example in this chapter of the Palletiser Chain Conveyor is a fairly simple machine and was used because it is (hopefully) quite easy to understand.

But even this simple machine has 9 MSIs leading to 5 Planned Maintenance schedules that include 14 SOPs.

MSIs like chain's, sprockets, bearings and Motor & Gearboxes occur very frequently across our breweries, the advantage of using a Steering Committee to steer the working groups developing the HACS system is that the Planned Maintenance schedules and Job Plans developed for these MSIs can be quickly copied the next time the same MSIs appear on another machine, or even on the same machine.

This is a critical point that allowed us to develop the HACS at speed and with consistency and highlights the key role of the Steering Committee. If you try to develop a Planned Maintenance system independently, for example at different breweries as was envisaged in the TPM Pillar route, then you will waste thousands of man hours developing Planned Maintenance schedules for the same MSIs as have already been developed elsewhere.

It was instrumental to have a steering committee that evaluated the work of the different students and Engineers working on different machines and made sure that they were not developing Planned Maintenance schedules that already existed on other machines.

This way of working is another reason why it is essential to complete the hierarchy 100% before starting to develop the Planned Maintenance schedules. Once the hierarchy is complete, it is easy to see which Planned Maintenance schedules can be copied within the hierarchy, or from other machines.

Once this way of working was established for the canning line HACS, I realised that it was going to be possible to "fix Planned Maintenance" in Asia and indeed all of HEINEKEN, if we continued with the next areas, building on the Planned Maintenance schedules that we had already developed for the canning line.

The more we developed, the easier the next machines became.

I said that in the beginning I only knew that I could develop the HACS for the SCALE canning line, and later I would worry about the rest of our equipment. This turned out to be a good strategy, because only once we had made good progress with the SCALE canning lines did it become clear how we would be able to attack the next areas, brewhouses and other packaging lines.

Developing a set of Planned Maintenance schedules is a process that requires a lot of discipline and a structured approach, and there is much more to it than just carrying out FMEA.

Once you have identified Failure Modes, whatever approach you use, there is a lot of work needed to select the correct maintenance strategy, and to then develop the Planned Maintenance schedule that will apply that maintenance strategy to the MSI. Frequently Engineers write Job Plans without specific standards, which are of very limited value. It takes a lot of work to describe clearly the standard and to put it into the Job Plan. Then even more work is needed to write the SOPs that are needed to describe the task when it is complex, and even when that is done we still need to upload the spare part numbers for all of the spare parts needed for each task as well.

When all that is finished, we need to verify what we have developed with suppliers or other independent experts, and we need some sort of database of the Planned Maintenance schedules as well as a governance system to collect modifications and make any changes that are needed.

Whilst many authors focus on the FMEA process, I have found that in building a really effective **Planned Maintenance** system that it is not in the FMEA that the real work lies. The real work is in the discipline to create Planned Maintenance schedules with specific Job Plans, clear SOPs and all other information built into the schedule.

I have included some sample Planned Maintenance schedules in Annexure Two for some of the more common items we have in the brewery, such as valves and pumps.

As a summary, here is a step-by-step guide to building a Planned Maintenance system that is hopefully explained clearly in this and the previous chapter:

13.5 HOW TO BUILD A PLANNED MAINTENANCE SYSTEM

- Develop the Hierarchy
    - Obtain all the documentation you can find for the machine from the supplier, especially the Spare Parts Catalogue.
    - Make sure that a structured asset numbering system is in place.
    - Use a standardised format for the Hierarchy.
    - Standardise the numbering of the machine sections.
    - Break the machine into sections and then into construction groups, then assemblies and sub-assemblies, then continue to MSI level.
    - Include only MSIs and not the whole BOM.
    - Every MSI is linked to an assembly, construction group and machine section.
    - Separate out the Field electrical in each construction group.
- Check and recheck that the Hierarchy is well structured, logical and covers the whole machine.
- Develop the Planned Maintenance schedules.
    - Form a working group to develop the Planned Maintenance schedules with machine experts (copy Planned Maintenance schedules that have already been developed for the same MSI elsewhere).
    - Identify likely failure modes for each MSI.
    - Work through the Maintenance task Decision Flowchart for each failure mode and identify the Maintenance Strategy.
    - Write a Planned Maintenance schedule for that Maintenance strategy.
    - Include lubrication and Calibration.
    - Define the assumed capability of the technician you are writing the job plan for.
    - Write a Job plan for each Planned Maintenance schedule. Check if other Planned Maintenance schedules should be included.

- Where the task is complex, develop an SOP to describe the task.
- List the SOPs and spare parts needed.
  - Use the upload sheet format for the Planned Maintenance schedule and the Job Plan for easier uploading of the data.
- Write the SOPs and make them available to Technicians.
- Verify the Planned Maintenance schedules and Job Plans that you are not 100% sure about with suppliers or other experts.
- Repeat for the next machine.

In developing the Planned Maintenance schedules, it is easier to complete the process for each MSI in turn consisting of identifying Failure Modes, selecting the Maintenance strategy, writing the Planned Maintenance schedule and the Job Plan before starting on the next MSI. Identifying all the failure modes for all the MSIs then all the maintenance strategies for all MSIs etc. means you are having to refocus on different components and is far less efficient.

That said, once the PM is complete, it is often the case that the SOPs are written later, when there is an opportunity to dismantle the relevant assembly (next section).

Clifford Jones

CHAPTER FOURTEEN: WRITING THE SOPs

As mentioned in the previous Chapter, the Planned Maintenance schedule is the WHAT and the WHEN of the Maintenance Strategy, and the Job Plan is the HOW.

The SOP is a more detailed and visual explanation of the HOW, required because our CMMS, Maximo, can only accept text in the Job Plan section and is not able to display Figures, pictures or videos.

As mentioned previously in Section 13.1, we decided that we will write all of our job plans assuming that we had a trained and qualified technician (he/she is able to use tools, take measurements etc.) but he/she is not familiar with the machine. So, when there is a maintenance task that requires complex disassembly or set-up, alignment or calibration, then we will need to have an SOP linked to the Job Plan.

SOPs can be a PDF, Word document or other format, but with any physically printed SOP there is an inherent risk that if the Technician is already at the machine, he/she may well try to complete the task unaided rather than return to the workshop or Planned Maintenance office to get the SOP to guide them.

We have investigated with Gebo/Sidel having the SOPs visible using an AR (Augmented Reality) headset, so that the SOP can be seen in a portion of the glasses that the Technician wears when carrying out the maintenance task. The level of detail and visual assistance of the AR solution was very good for an initial training activity but becomes excessive and is mostly ignored once the Technician has completed the task a few times. We felt that the hours needed to develop each set of AR instructions was therefore not justified (3D models of the parts are required), and that the AR visualization did not add much value when compared to having a good SOP on a tablet or other mobile device. We concluded that a simple graphic SOP was a better trade off in terms of the required development time.

Swipeguide is an intuitive tool in which to build an SOP and is able to handle a large amount of pictures and video, plus it is easy for the Technician to view each instruction in sequence.

Therefore, our SOPs are made using Swipeguide, which has the advantage that they are available on a mobile device such as the handphone of the Technician. It is thus convenient for him/her to access the SOP, as it will almost always be the case that the Technician has their mobile phone with them. (Swipeguide, 2013): https://www.swipeguide.com/

At the time of writing the HACS system contains nearly 10 000 individual Planned Maintenance schedules, and for those there are nearly 900 SOPs in Swipeguide.

In order to write the SOPs, we assign one of the Working Group to work with a Technician on the machine and to photograph each step of the process that we are trying to explain. To make the SOP we use the Swipeguide Editor that allows us to upload each photo and enter the necessary text to describe the steps.

14.1 GENERIC SOPs

The Steering Committee realized that there are "generic" tasks that we need to explain, such as how to measure chain elongation (INSPECTION), how to replace a seal in a valve (REPLACEMENT), how to change the oil in a gearbox (LUBRICATION), or how to calibrate a temperature probe (CALIBRATION). We did not want to repeat these SOPs for every machine that had the same MSI, so we made a set of GENERIC SOPs that can be applied to any machine that includes that MSI.

An SOP was categorized as generic when the MSI was likely to be found on more than one machine.

Generic SOP's have only a 6 Character identifier in our Swipeguide files, the first 3 characters is either INS, REP, LUB or CAL, for Inspection, Replacement,

Lubrication or Calibration, and the next 3 characters are a sequential number for that maintenance activity type.
Examples:

- INS-003
- LUB-2
- REP-12
- CAL-004

EXAMPLE OF GENERIC INSPECTION SOP:

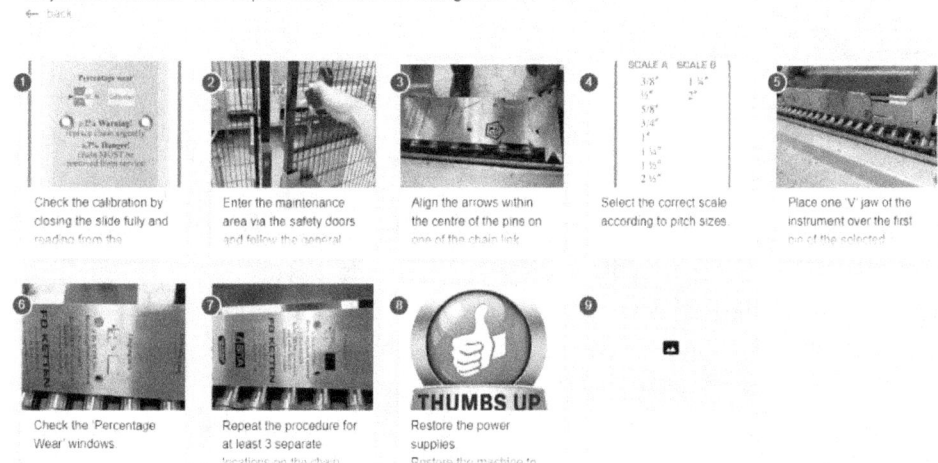

Figure 82: Generic Inspection SOP from Swipeguide.

Figure 82 shows the Swipeguide SOP to inspect chain elongation using the FB Ketten chain stretch gauge. This SOP only needs 9 steps to show the Technician how to make the measurement and how to read the result on the gauge.

Do note that only part of the text of each step is shown in the summary views that I have used here from the Swipeguide Editor.

What the Technician sees on his mobile device is each single image with the text below, and when he swipes across the screen he sees the next image with its associated text.

EXAMPLE OF GENERIC REPLACEMENT SOP:

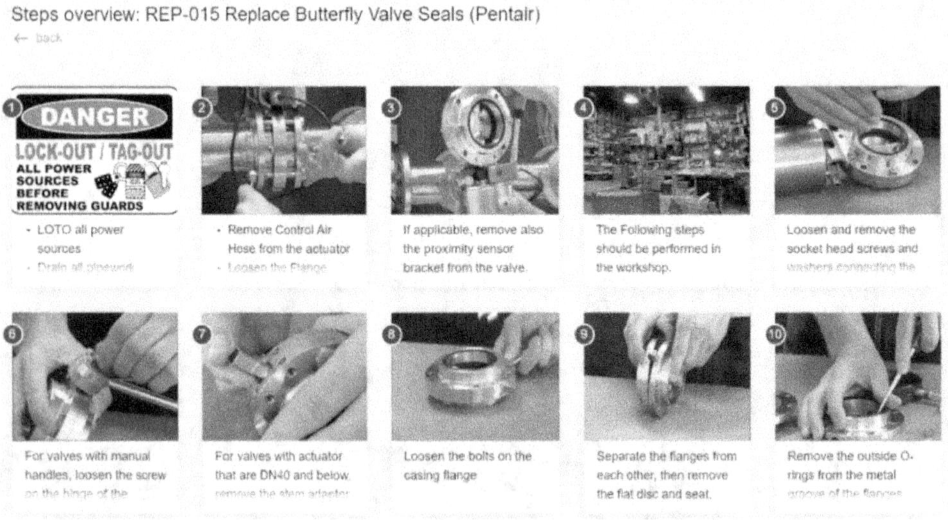

Steps overview: REP-015 Replace Butterfly Valve Seals (Pentair)
← back

Figure 83: SOP to replace butterfly valve seal from Swipeguide.

Figure 83 is the SOP to change the seal in a butterfly valve, of which there will be several hundred in a typical brewery. (This is a Time-Based Replacement activity because the seal is very cheap, refer to Chapter 8).

The complete SOP is over 30 steps, only some of them are shown here.

YouTube contains many videos from diverse sources of maintenance activities, many suppliers have posted videos of how to carry out maintenance tasks for their equipment. In this case the images are screen captures that we took from a YouTube video of how to change the valve seal.

(Step 4 is not our workshop in such a bad state, but it shows that the Engineer writing the SOP has a sense of humour).

In Chapter 13 I said that we had decided not to describe maintenance tasks carried out in the workshop or off-line, and only to describe those that are carried out on the production line itself. There are some exceptions to this rule, arising if a task is likely to be carried out frequently by our own Technicians (rather than a 3rd party supplier), and if it was a straightforward matter to produce the SOP, as it was for the butterfly valve seal.

EXAMPLE OF GENERIC LUBRICATION SOP:

Steps overview: LUB-001 Lubrication of Motor Gearbox
← back

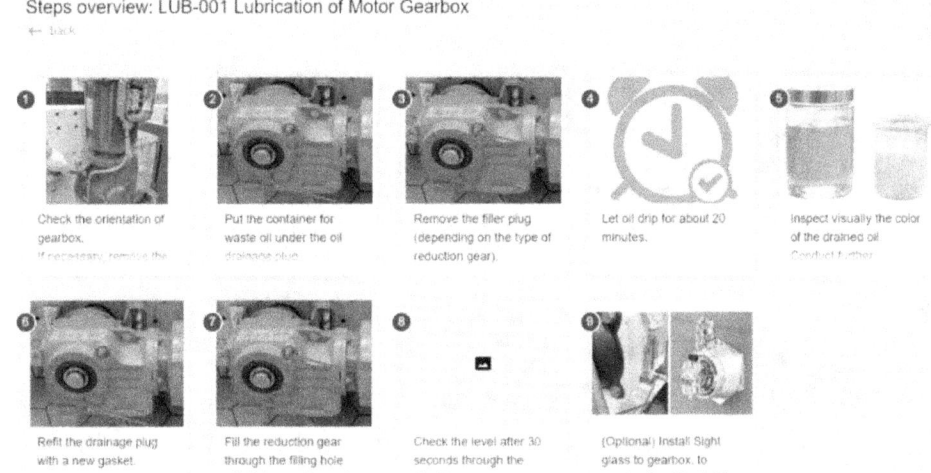

Figure 84: SOP to change oil in a Motor Gearbox from Swipeguide.

Figure 84 shows how to change the oil in a gearbox. The inspection of the oil for water presence is important.

EXAMPLE OF GENERIC CALIBRATION SOP:

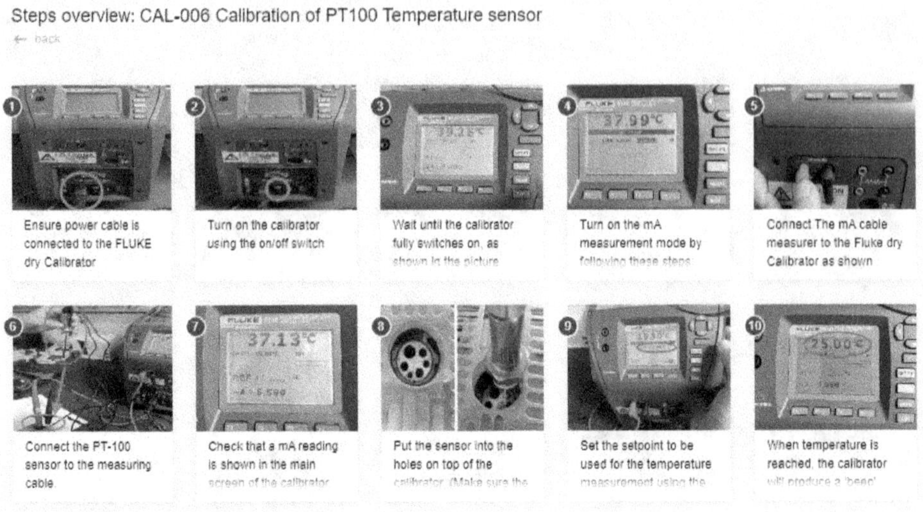

Steps overview: CAL-006 Calibration of PT100 Temperature sensor
← back

Figure 85: SOP to calibrate a temperature probe from Swipeguide.

Figure 85 is a part of the SOP to calibrate a temperature sensor, referred to as a PT-100 (Platinum, resistance of 100 Ohms at 0°C). The importance of this type of sensor is explained in section 15.4

For this application we chose the Fluke dry-well calibrator, but there are other types of equipment available. This led us to develop a list of all of the correct tools needed to carry out the Planned Maintenance schedules described in Chapter 15.

14.2 EXAMPLE OF MACHINE SPECIFIC SOP

For each machine there will usually be some specific maintenance activities that need an SOP to clarify the steps of the Job Plan.

For machine specific SOPs the SOP is given a longer identifier in the format: XX-YY-ZZ-123 where XX identifies the supplier, YY the area, ZZ the machine and then a number sequence.

Therefore: GC-CL-PL-006 is the SOP for Gebo Cermex-Canning Line – Palletiser – 006.

Steps overview: GC-CL-PL-006 Replacement of Chains and Chain Guide Wheels for Opening Roller

← back

① Lower the Layer Deposit Module to suitable height for working.	② Enter the maintenance area via the safety doors and follow the general	③ Push the side bars to the center.	④ Mark the positions of movable pulleys for both sides.	⑤ Release tension of both toothed belts by adjusting the tensioners at the
⑥ Unscrew and detach opening roller carriage from the toothed belt by	⑦ Unscrew and detach opening roller carriage from the toothed belt by	⑧ Remove both carriages from top openings of opening roller guides.	⑨ Unscrew all screws to detach the carriage rollers from roller opening chains	⑩ Replace with new roller openings chains and chain guide wheels
⑪ Slide opening roller carriages back into opening roller guides	⑫ Screw and connect opening roller carriages back to toothed belt.	⑬ Adjust tension of the toothed belt back to marked position, or to the	⑭ Restore the power supplies. Set the machine into	⑮ THUMBS UP Restore the machine to AUTOMATIC MODE.

Figure 86: SOP for Sidel Palletiser from Swipeguide.

Figure 86 shows the SOP to replace the chains and chain guide wheels on the palletiser Layer Deposit Module. It should be clear that this SOP is very specific to the design of this machine. Other machines will not have this Layer Deposit Module and so to illustrate how to change the chains and chain guides we make a machine specific SOP.

As I explained in the previous chapter, developing good Planned Maintenance schedules is a lot more complex than just carrying out an FMEA. We have to build the asset hierarchy, identifying every MSI logically in terms of asset, construction group and assembly. Once that is complete we try to identify the typical failure modes for that MSI, and then use the Maintenance Task Decision Flowchart to identify the Maintenance Strategy. Next we write the Planned Maintenance schedule, describing the WHAT and the WHEN of the maintenance tasks, and then we can write the Job Plan (the HOW), including a standard for all inspections. This chapter describes only one part of that Job Plan, the SOP that we will write where we need an illustration or picture to show the Technician HOW to carry out the maintenance task.

CHAPTER FIFTEEN: TOOLS AND CALIBRATION

As we started to write inspection schedules for the HACS, we realized that we also needed to standardize the tools used for those Inspections and Calibrations. The HACS is designed to be copied to as many of our approximately 160 breweries worldwide as possible, so it is important that everyone has the same tools so that the steps in the Job Plan and/or SOP can be followed easily.

Included in the HACS database is a list of tools that should be used to carry out the HACS schedules. The following are some of the most important tools used in the HACS:

15.1 GENERAL TOOLS

When we issue a Planned Maintenance schedule it will refer to the (TAG) number of the asset, and for items like electric motors, isolator switches and photocells etc.. they need to be clearly numbered so that the Technician can find the correct one.

One of the best ways to do this is with a plastic engraved label, which can be produced in house with a pantographic engraving machine, shown in Figure 87. Of course, there are more expensive electronic engravers available, but for asset numbering this is all that is really needed. The plastic labels are then fixed next to the item, often on the conveyor frame. I have found the engraved labels to be very durable, whereas printed labels eventually fall off due to environmental conditions.

It is important to have the engraving machine available in-house as this makes it convenient and easy to quickly produce the necessary labels. If you use an outside supplier to make labels it can take several weeks to complete the procurement process for each batch, and this slows down progress.

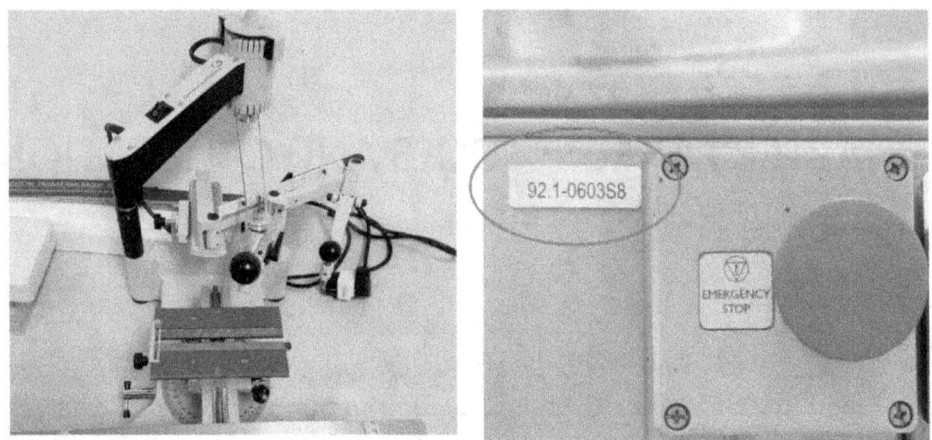

Figure 87: Pantographic Engraving machine and engraved asset tag number.

In the HACS we also specify which tools the Technician should have in his/her toolbox. The full list is included in Annexure 5.

15.2 LUBRICATION TOOLS

ULTRAPROBE 401 GREASE CADDY:

Figure 88: Ultraprobe 401 grease caddy, from author's collection.

The Ultraprobe 401 grease caddy shown in Figure 88 is a lubrication tool (from the same company as the ultrasonic leak detector in the following section), (Grease Caddy, 2023) that overcomes the challenge of how to ensure the correct amount of grease is applied to a bearing (not too much, not too little) by using an ultrasonic detector to signal when the bearing is correctly lubricated as the grease is being applied. This is described in more detail in Chapter 16, Lubrication.

https://www.uesystems.com/product/ultraprobe-401-digital-grease-caddy-pro/

15.3 INSPECTION TOOLS

F B KETTEN CHAIN STRETCH GAUGE:

Figure 89: Technician using a F B Ketten chain stretch gauge, from author's collection.

The FB Ketten chain stretch gauge shown in Figure 89, https://www.fb-ketten.de/products/verschlei%C3%9Fmesslehre , (Chain Stretch Gauge, 2022) is one of several similar gauges available from chain suppliers. We found this one to be easy to use. It includes a gauge to measure the chain pitch, it indicates how many links to check, and the percentage stretch is shown in a window on the gauge. The tool makes inspection of chain stretch very fast and accurate. The Generic SOP for using this tool is included in section 14.1.

UE ULTRAPROBE ULTRASONIC LEAK DETECTOR:

Figure 90: UE Ultraprobe ultrasonic leak detector, from author's own collection.

Breweries have many valves and pistons operated by compressed air. The air is supplied to the devices via flexible pipes that fit into couplings that have a considerable tendency to leak. This leakage causes a great deal of wasted energy at the air compressors and can cause a failure of the valve or piston to operate. Because of the noise level in a brewery, small air leaks are very difficult to detect. UE systems sell a range of ultrasonic leak detectors that can identify air leaks (or other gasses) using ultrasound and can be used during normal operations, the background noise level does not affect the device. The ultrasound is converted to an audible signal that the operator hears in his headset.

We use the Ultraprobe 3000 (Ultrasonic leak detector, 2023) for leak detection in the brewery https://www.uesystems.com/applications/ultrasonic-leak-detection/ , shown in Figure 90 where a Technician is checking for a leak on a steam control valve.

Recently we have started to use the Fluke i900 imager (Sonic industrial imager, 2023) that has the advantage of giving a visual picture and showing the leak on the screen of the device. https://www.fluke.com/en/product/industrial-imaging/sonic-industrial-imager-ii900

CHECKLINE BTM400 BELT TENSION FREQUENCY METER:

Figure 91: Technician using Checkline BTM400+ to check belt tension.

We described in Chapter 13 how we check the tension on a chain drive, the total mid-span movement in the slack span of the chain should be approximately 5% of the distance between the driving and driven sprockets.

For a toothed belt it is more difficult to set the correct tension because the belt is quite flexible, and this can lead to very different interpretations between Technicians as to what is the correct amount of tension.

An innovative solution is available from Checkline (Frequency Gauge, 2023), who manufacture a belt frequency gauge, shown in use in Figure 91. You strike the belt with a suitable tool, such as the head of a screwdriver, and the probe on the device measures the frequency of vibration of the belt and can convert this to tension in Newtons. We have to have a different standard for nearly every belt, but the inspection is very fast, accurate and consistent with this tool (which can be used for any type of belt): https://www.checkline.com/product/BTM-400PLUS

GAPPSCAN G2 LEAK DETECTOR:

Figure 92: Brewery Plate Heat Exchanger

Plate Heat Exchangers (PHE) as shown in Figure 92 are used in several locations in the brewery, often to cool our product at a critical stage of the process, such as before fermentation. These devices have many thin plates containing channels which alternatively carry the product being cooled and the media doing the cooling. The plates are sealed with gaskets and are designed to be as thin as possible in order to facilitate heat transfer.

Failure modes of the PHE include leaking gaskets and microscopic holes or cracks in the plates. If the cooling media and product mix, it can lead to an expensive product recall if contaminated product is not detected before sale.

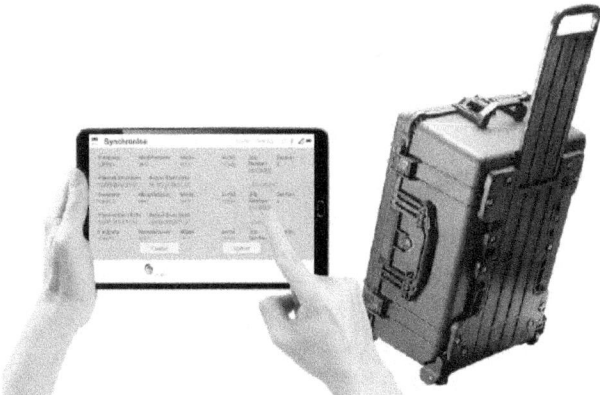

Figure 93: Gappscan G2 for leak detection in Plate Heat Exchangers (PHE)

Conventional pressure testing of the PHE is not sufficiently accurate to detect very small leaks, so the maintenance strategy before the HACS was developed was to carry out a complete tear down overhaul of the PHE's periodically and then use a penetrating dye brushed onto the plates to detect cracks or pinholes, and then to reassemble the PHE with new gaskets.

This process had a high risk of cracks not being detected as the penetrating dye requires visual inspection, and the dye has to be completely removed before the PHE can be put back into use. The gasket sets are very expensive and there is a significant risk of a leak occurring from the re-assembly, or of damage occurring that would not have occurred if the plates had been left alone (infant mortality failure).

From 2019 I started working with EIT to test and validate the Gappscan leak detector (Gappscan, 2022), Figure 93, testing the device in several breweries and supervising an SIT (Singapore Institute of Technology) student to carry out an assessment of the reliability and repeatability of the device. (This is known as R&R and is a statistical assessment of the capabilities of a gauge).

We were able to show that the device is very accurate in detecting leaks down to the size of a few microns. The maintenance strategy for PHE's is now to inspect every 6 months with the device, to monitor any very small leaks (below 25micron we do not take action) and to carry out a teardown and inspection only once there is a trend or significant issue.

Our validation process led to some modifications being incorporated into the latest G2 model, which is now in use in several of our breweries and is saving tens or hundreds of thousands of dollars in unnecessary teardown overhauls of PHE's.

https://www.eit-international.com/gappscan-flyer/

OTHER INSPECTION TOOLS

Other tools required to complete the HACS inspection schedules are listed below, but here the SOP is such that the particular device does not have to be standardised, so it is possible to use the recommended model or an alternative.:

- Dial Gauge.
- Laser distance measure (Bosch GLM).
- Laser shaft alignment tool (recommended Fluke 830).
- Thermal imaging camera (recommended Fluke FLIR E53 24°).
- Infrared gun type thermometer (recommended Fluke IR thermometer/568 62Max).
- Plastic conveyor elongation measurement tape: Often available from the conveyor chain provider, such as System Plast https://www.regalrexnord.com/brands/system-plast
- Gauss meter/Tessla meter (recommended Goudsmit Gaussmeter HGM09).

15.4 CALIBRATION TOOLS

E+H PROLINE FIELDCHECK SIMULATOR:

Figure 94: Endress+Hauser field-check simulator

Endress + Hauser produce an extensive range of flowmeters and other devices such as conductivity meters, level sensors etc..: (Field Instruments overview, 2023) https://www.endress.com/en/field-instruments-overview

Their flowmeters are found in many breweries and beverage plants.

Calibration of a flow meter requires that it is installed in a calibration test rig where a known volume of liquid is passed through the flowmeter and the reading is compared with that of either a standard meter or it is measured in another way, such as using a calibrated collection tank. Such test rigs are expensive and rarely available at our breweries, generally our flow meters have to be sent to a specialized company to have them calibrated.

The Endress + Hauser Proline Field-Check™ Flow Simulator is a very useful VERIFICATION device, shown in Figure 94, it connects to an E+H flowmeter and simulates liquid flow, and is able to verify the correct operation of the sensors and circuits in the flowmeter. In the HACS we specify verification for flow meters supported by less frequent calibration.

Endress + Hauser have phased out the Field-Check Flow Simulator as it is being replaced by their Heartbeat system that verifies flowmeters through a network connection. The Heartbeat system only works on their newer flow meters, the Field-Check simulator is still very useful for the older flowmeters.

FLUKE 9142 FIELD METROLOGY WELL:

Figure 95: Fluke 9142 Field Metrology Well, from www.Fluke.com

Figure 95 shows the Fluke 9142 Field Metrology well used to calibrate PT-100 (Platinum, resistance of 100 Ohms at 0°C) temperature sensors.

These sensors are found all over a brewery and are particularly important in the fermentation tanks where they measure the temperature of the beer and via a PLC they control the amount of coolant flowing to the fermenter tank's cooling jackets. There are many other points in the brewery process that accurate temperature control is important, such as in the pasteurizer in the Packaging process.

If these temperature probes are not accurately measuring the true temperature, we can have high energy consumption and/or process variations that affect product quality.

When we started building the HACS many breweries used a calibrated temperature bath in the laboratory to check if the sensor indicated the correct temperature of the bath, but this does not provide a full calibration of the sensor across its operating range. The Fluke Field Metrology well allows the signal in mA from the sensor to be checked at a range of temperatures across its range so that the sensor can be adjusted if necessary to achieve the correct calibration.

FLUKE 718 PRESSURE TRANSMITTER CALIBRATOR:

Figure 96: Fluke 718 300G Pressure transmitter calibrator from www.fluke.com

There are many pressure sensors in the brewery that transmit a signal to the control system. These are important as they control the liquid or gas pressure in our piping and pumping systems.

The Fluke pressure transmitter Calibrator in Figure 96 is a handheld device that is able to receive a pneumatic and electrical signal to compare and calibrate a pressure sensor.

FLUKE P5510 COMPARITOR:

Figure 97: Fluke P5510 comparitor from www.fluke.com

Calibration of pressure gauges and safety valves was not carried out consistently in our breweries, some breweries used third parties, some used various different devices, and some did not have any effective in-house facility.

The Fluke comparator is used to calibrate both pressure gauges and to verify the operation of pressure safety valves.

To calibrate mechanical pressure gauges, the gauge that is to be calibrated is fitted into the comparator shown in Figure 97, next to a standard (certified) pressure gauge, such as a Fluke 2700G BG2M reference gauge. The comparator is then hand pumped to different pressures and the reading on the gauge to be checked is compared with the reading on the reference gauge. Not all dial pressure gauges have a calibration adjustment screw, some have a screw that only allows you to adjust the zero point on the gauge, others have a screw that can adjust the reading at a certain pressure. Whatever the type of gauge, the comparator can at least be used to check that the gauge is giving an accurate reading, and if it is not, and it cannot be adjusted, then it can be replaced.

As we developed the HACS we found that there are some pressure relief valves in our breweries that are not regularly tested for safe operation. The pressure relief valves on our boilers and steam system are of course regularly tested due to pressure certification requirements, as are those on any pressure vessels such as CO_2 storage tanks. But there are many pressure relief valves in our ammonia piping and CO_2 piping that were never or very infrequently tested for correct operation.

As we built the HACS for each machine and process we made sure that every pressure relief valve had a Calibration Planned Maintenance schedule included. Actually, it is not a "calibration", as in this case we do not want the technician to adjust the valve. The Planned Maintenance schedule Job Plan instructs the technician to put the pressure relief valve onto the comparator and check that the valve opens at the correct pressure, and if not to replace it with a new one. What we do is actually a verification of the operation of the safety valve.

The Planned Maintenance schedule is labelled as a Calibration, but the task is to verify the correct operation of the safety valve.

This step has made a significant improvement in process safety in our breweries, and I hope that this will be copied and implemented in other companies that may be in the same situation.

Clifford Jones

CHAPTER SIXTEEN: LUBRICATION

16.1 GENERAL LUBRICATION PRINCIPLES

Lubrication is a key maintenance task whereby we apply a lubricant to reduce the friction and wear that occurs when there is contact between 2 surfaces that are moving relative to each other. The lubricant forms a film between the 2 surfaces.

In our breweries the lubrication tasks are carried out by Engineering Technicians at the foundation organisational level, but we try to transfer the tasks to the Operator via Autonomous Maintenance as we move to the more advanced organisational level.

As we build the HACS we look at each MSI and how it might fail, as well as what are its lubrication requirements if it is a moving part or if it contains moving parts. Usually, this requirement is clearly specified by the manufacturer, the type of lubricant and the frequency of application is given in even the most basic maintenance manuals.

When we transfer the lubrication task to operators it is still issued as a Planned Maintenance schedule with a Job Plan, but the Operator becomes the responsible person. Issuing the Planned Maintenance schedule is important because every Planned Maintenance schedule is tracked for completion and the responsible person has to sign and confirm that the task has been completed, as well as comment on any issues or difficulties.

For each machine we need to make a lubrication map as shown in Figure 98 which shows the location, frequency of and responsibility for the lubrication.

AM 01 - MAP OF ZONE OF LUBRICATION FOR FILLER-SEAMER-RINSER-CONVEYOR

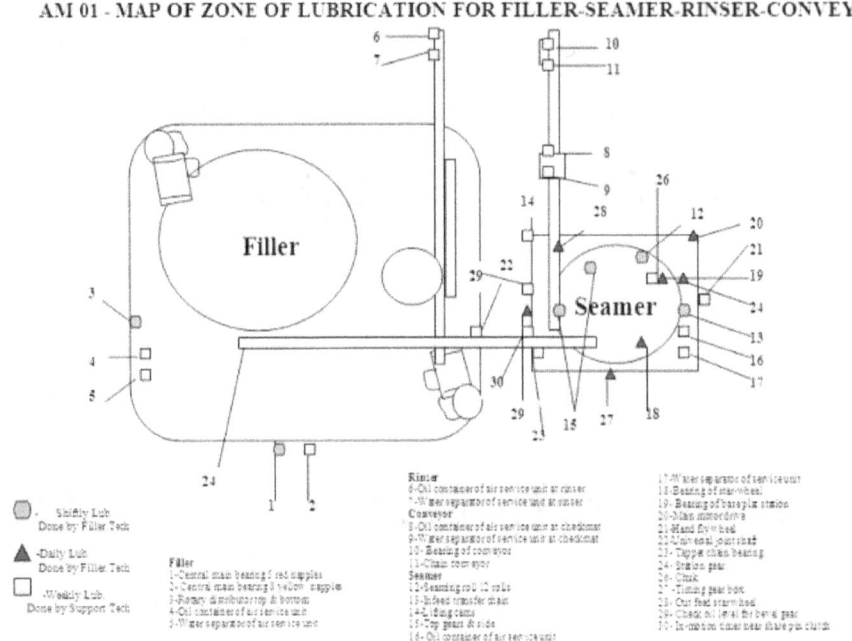

Figure 98: Filler and seamer lubrication chart, HEINEKEN lubrication manual

At each lubrication point we need to give a clear indication of the lubricant to be applied to make sure that we are not using the wrong lubricant. To do this we usually put a symbol at the lubrication point, and also the same symbol on the grease gun or oil dispenser and on the oil or grease storage drum.

The problem is that there could be 30 or even 40 different lubricants in use in a brewery, and it is hard to find that many different colours or shapes for clear identification. The reason we have so many lubricants is that we have some equipment that requires special lubricants, such as in the cooling compressors, and we have some equipment where there can be contact with the product, so there we need to use a food grade lubricant.

A further complication is often that one equipment manufacturer recommends a lubricant from a particular supplier, and a different equipment manufacturer recommends a lubricant from a different supplier, but they may both be fundamentally the same lubricant (such as a grease or oil) with different brand names. In order to make sure we do not invalidate guarantees on the equipment, we have to use the specified lubricant and we end up with an unnecessary number of sku's in the lubrication store. The only way to address this is to standardize as much as possible on one lubricant brand, and then try to get machine manufacturers to confirm in writing that the lubricants of that brand can be used on their machine.

Item	Shell	Color code	Main application	Remark	Item	Shell	Color code	Main application	Remark	Item	Shell	Color code	Main application	Remark
1	SHELL TELLUS 46		Hydraulic system, forklift truck / Hydraulic cylinder (malt/rice lifting platform) / Lifting cylinder - Bottle filler	Oil	11	SHELL OMALA 460		Gear box - Clarifier - Can filler / Bottle pasteuriser, Bottle filler, Gear box of Lauter Tun	Oil	21	SHELL TONA T 68	△	Workshop - Milling machine	Oil
2	SHELL CLAVUS SD 22.12		Refrigerational machine (FX-10)	Oil	12	SHELL EDM Fluid 2		Hidrostal - Submersible pump	Oil	22	Kluberquiet BHQ 72-102	▲	Motor bearing of NH3 compressor S3	Grease
3	SHELL CORENA S 68	●	Screw air compressor - Kieselguhr dosing pump	Oil	13	SHELL TIVELA S 460	▨	Worm gear boxes of Krones	Synthetic oil	23	Molykote	△	Motor bearing of NH3 compressor S4	Grease
4	SHELL RIMULA 15 W - 40	●	Diesel engine, pump, general purpose oil	Oil	14	SHELL CASSIDA HF 100	■	Can seamer	Food grade	24	Sacoke grease		Burner Sacoke	Grease
5	SHELL CORENA P 100		Piston air compressor, Seeger compressor	Oil	15	SHELL TIVELA S 220		Geared motors of can/bottle conveyors, Bottle washer / Gear boxes of Bottling line	Synthetic oil	25	Kluber lubrication Staburags NBU12	△	Air blower	Grease
6	SHELL ALVANIA EP 0		Conveyor drive (Malt & Rice)	Industrial grease	16	SHELL MORLINA 10		Air lubrication system of Krones, EBI, Labeller / Air service unit of Canning & Bottling line	Oil	26	Sabroe Pao 68	◉	NH3 comp. package No2	Oil
7	SHELL CASSIDA CHAIN OIL 1000		EBI, Bottle filler	Food oil	17	GENUINE HOSE LUBRICANT		Hose pump	Food grade	27				
8	SHELL RETINAX LX 2	△	Single auto (EBI &) / All bearings, general purpose / Big cap motors	Industrial grease	18	SHELL CLAVUS 68	⬡	NH3 compressor	Oil	28				
9	SHELL CASSIDA EPS 1	▲	Rotary distributor of Can filler / Bottle filler (roller of crowning head) / Lauter Tun, Mash Tun, Mash Copper, Malt & Rice Equipmet / Single lubrication point, ECI conveyor	Food grease	19	SHELL OMALA 680	⬟	Worm gear boxes, Speed reducer	Oil	29				
10	SHELL TONA T 32	△	Workshop - Milling machine	Oil	20	SHELL OMALA 220	⬡	Transmission roller chain Helical, bevel gear boxes speed reducer of labeller	Oil	30				

Figure 99: Lubrication identification system, HEINEKEN lubrication manual

Figure 99 shows a possible lubrication identification system where a combination of shapes and colours are used to identify the different lubricants used in different locations (This example focusing mostly on Shell as the preferred brand).

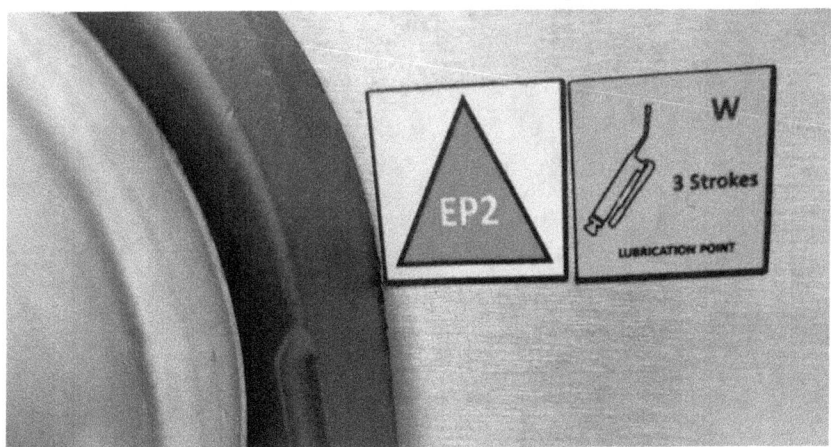

Figure 100: Lubrication point identification, author's own collection.

In Figure 100 we can see a well-designed lubrication point identification symbol.

The red triangle with letters EP2 identifies the grease to be used, and we should find the same red triangle on the grease gun and the grease supply drum in the store to prevent any confusion.

The second green sticker has a letter W which indicates that the lubrication is done WEEKLY.

Underneath that, we see 3 STROKES, meaning that the Technician or Operator should apply 3 strokes of the grease gun pump handle to apply the correct grease amount. There is additionally a symbol of a grease gun to be really clear as to which tool is required here.

The specification of 3 STROKES is an approximation of the correct amount of grease based on experience. In the next section we will look more closely at the use of the Ultraprobe Grease Caddy to control the amount of grease applied.

16.2 LUBRICATION SET UP

Figure 101: Lubrication set up at TAP brewery in Thailand, author's own collection.

One of the best arrangements of lubrication storage and distribution that I have seen has been carried out by Pakorn Sumritpol and his team at the Thai Asia Pacific Brewery (they also have excellent spare parts store organization, see Chapter 18), some of the main points are shown in Figure 101.

Oil storage drums are on spill pallets and have a locally designed float system to indicate the remaining contents using 3D printed parts (This is important for stock taking and knowing when the drum level is low).

A local lubrication station is installed in each area where it is needed, such as Packaging, Utilities and Brewing areas, and it contains the lubricants needed in that area for ease of access.

Oil pumps and containers are clearly labelled and standardised, and the storage of specialized items used in small quantities, such as spray lubricants, is well organized.

16.3 ULTRAPROBE GREASE CADDY OPERATION

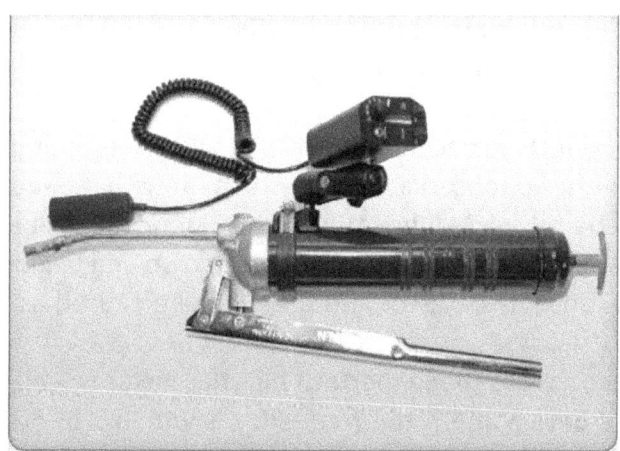

Figure 102: Ultraprobe 401 grease caddy, from uesystems.com

As mentioned in Chapter 15, the Ultraprobe 401 grease caddy (Grease Caddy, 2023) is a lubrication tool from UE systems, that overcomes the challenge of how to ensure the correct amount of grease is applied to a bearing (not too much, not too little) by using an ultrasonic detector to signal when the bearing is correctly lubricated as the grease is being applied (Figure 102).

https://www.uesystems.com/product/ultraprobe-401-digital-grease-caddy-pro/

Ultrasound instruments in this application detect changes related to friction. A properly lubricated bearing will have very little friction. The lubricant evens out any stress that the bearing encounters as it rolls around the raceway.

As the lubrication level in a bearing falls, the potential for friction increases. There will be a corresponding rise in the ultrasound amplitude level that can be detected.

There are two potential failure modes of a bearing relating to lubrication: lack of lubrication and over lubrication.

In the presence of a lubricant film on a bearing surface, there is a dampening effect of the stresses caused by microscopic imperfections, and the acoustic energy produced will be low. Should lubrication be reduced to a point where the stress damping is no longer present, the rough spots will make contact with the bearing surfaces and increase the acoustic energy. This will begin to produce wear and small fissures may develop. The service life of a bearing is strongly influenced by the relative film thickness provided by the lubricant.

When too much lubricant is put into the bearing housing, pressure builds up and can lead to an increase of heat, which can create stress and deformity of the bearing.

The baseline for a bearing reflects in decibels the level at which it is operating under normal conditions with no observable defects and with adequate lubrication. The Baseline can be set while lubricating. While lubrication is being applied the Technician will listen until the sound level drops down and begins to rise. At that point no more lubricant is added, and the dB value is used as the baseline.

After the baseline inspection has been performed, the bearing can be inspected with the ultrasonic device and if the frequency is within 8dB of the baseline then no more lubrication is needed. If the bearing exceeds 8 dB over a set baseline it needs further lubrication.

16.4 SEALED FOR LIFE BEARINGS

Currently we are evaluating Sealmaster bearings from System Plaast (Sealmaster, 2022), Figure 103, that are sealed for life and require no lubrication https://www.regalrexnord.com/brands/Sealmaster

Of course, the life of this bearing will be shorter than a standard bearing, 4 years is anticipated by the manufacturer.

The sealed for life bearing incorporates parts that collect the lubricant pushed out of the ball race and return it to the bearing surfaces so that the grease is recirculated instead of discharged.

Figure 103: Sealed for life conveyor bearing, from author's own collection.

CHAPTER SEVENTEEN: APPLYING THE HPO MODEL

The HPO (High Performing Organisation) framework and HPO model have been researched and published by Andre de Waal and David Hanna in their paper https://www.hpocenter.com/article/hpo-model-hpo-framework-organizational-improvement-for-a-european-multinational , (Hanna, 2016) and having read the paper I strongly suspect that I know which is the unnamed European Multinational that they refer to.

The HPO framework was developed according to a review of academic publications on high performance and a questionnaire that was completed by 3,200 respondents worldwide.

According to the framework, an HPO is an organization that achieves results that are significantly better than those of its peer group by focusing on what really matters to the organization.

The HPO framework consists of 5 HPO factors.

1. Management Quality.
2. Openness and Action-Orientation.
3. Long-Term Orientation.
4. Continuous Improvement and Renewal.
5. Employee Quality.

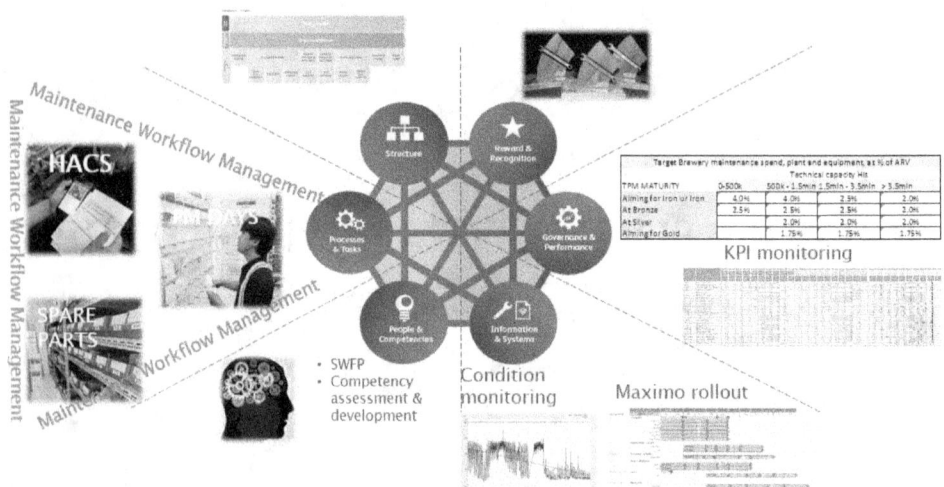

Figure 104: HPO model applied to Planned Maintenance, author's training materials.

De Waal's framework identifies WHAT companies need to do to become a High Performing Organisation, while Hanna's model indicates HOW companies can change their organizational elements to become HPOs.

Hanna's HPO model looks at 6 key organisational elements (design elements) that need to be adapted together to achieve a new organisational strategy.

As we built the HACS it became clear that to "Fix Planned Maintenance" (see Chapter 5) it would take more than just the HACS alone. For the first couple of years, I deliberately ignored this, as I needed to devote every resource I had just to developing the Planned Maintenance schedules.

But the HACS has to operate in an environment where we have well organized spare parts, Planned Maintenance days, capable people, the right organisational structure and resources etc. it was this environment that the Sahara project tried to address, and it is the combination of these factors that is needed to build an effective Planned Maintenance system.

Once we were making good progress on the HACS development, after the first two years or so, I had to start thinking more about the environment that the HACs operated in.

So, I applied the HPO model to Planned Maintenance in Asia Pacific, the model shown in Figure 104 is now applied by HEINEKEN globally. Much of the inspiration for this came from Jan Paul Boon, and in the processes and tasks I included Workflow Management, PM Days and Spare parts Management that had already been developed by Centre's of Excellence led by the Global Maintenance team.

17.1 THE SIX DESIGN ELEMENTS:

There are six key design elements in the HPO model:

PROCESSES AND TASKS: *We need to ask which processes and tasks have to be improved and in what way to make sure that we get real improvement?*

In our Planned Maintenance model there are 3 Processes and Tasks, The HACS which are of course the actual Planned Maintenance schedules, the Spare Parts (Chapter 18) and the Planned Maintenance days (Chapter 19). These 3 processes form what I call the TITANIUM TRIANGLE and they are connected by Maintenance Workflow management. (Why Titanium? because it is strong and light 😊).

In terms of Planned Maintenance, we can build an effective maintenance system if these 3 processes are well managed.

STRUCTURE: *Do we need to change or adapt the organisational structure to achieve better co-operation and alignment? Do we have to shift or change roles and responsibilities?*

Happily, the answer to these 2 questions is NO. Through the TPM process the optimum organisational structure for all departments including Engineering has been very well designed and is available for the different organisational maturities of Foundation, Advanced and Word Class. The specific numbers needed will change with the workload at the particular brewery, but the correct structure with required functions such as Maintenance Engineer or Planned Maintenance Planner is well defined in our company.

PEOPLE AND COMPETENCIES: *How can we strengthen the competencies that people need to be able to work in a simplified and aligned manner?*

Apart from needing to "Fix Planned Maintenance", one other area that we identified as an issue in the APAC region in 2017 was a lack of Strategic Workforce Planning (SWFP). Finding that each brewery had a different way of working, I built a framework for SWFP to support each site in the SWFP process.

The first step in SWFP is to identify the current maturity of the organization and the desired maturity level by department. Then to identify the structure needed. The biggest step is to conduct a functional competency assessment of each employee, to establish if that employee has the necessary competencies for his/her role (gap analysis), and to evaluate whether we think that employee will be able to develop to fit the future organisational level. To do this a detailed mapping of the Functional Competencies needed for each job role is required, which fortunately had been developed in previous years, it was a project that Jan Paul had been closely involved in. The key output from the gap analysis is a comprehensive training plan to fill the competency gap.

In order to close the gap, we have to have the necessary training staff and training facilities, which is why in 2021 we built a Training Centre in Cambodia (at relatively low cost by refurbishing existing buildings) and added a Training Manager and 2 Trainers to our structure.

One challenge from this process is that when the competency gaps are analysed it becomes very complex to decide what training to start with. A good approach is to identify the top 5 or so training needs in each function and address those, so that you reduce the competency gap for the maximum number of people.

In 2022 in Cambodia, we carried out 103 hours of training per employee, we think this is a good benchmark, but we also pay a lot of attention to the quality of the training, not just the quantity.

We certainly benefit directly from the increase in capability this training provides, but we also benefit in reduced turnover as employees see our company as a great place to work that foster's their development.

REWARD AND RECOGNITION: *In what way do we need to change the reward and recognition system so that it fosters instead of hinders co-operation and collaboration?*

In terms of Planned Maintenance or Engineering KPI's we did not have any Reward and Recognition system in place. I proposed the concept of the Golden Spanners Award and in 2022 the Global Maintenance team arranged the first Golden Spanners Awards with 6 categories to recognize the achievements of those working on Planned Maintenance:

- HACS development.
- Planned Maintenance Days.
- People development.
- Spare Parts.
- Cost Management.
- Improvement teams.

There were 75 nominations received from across the world, showing the enthusiasm and recognition that the Global Maintenance Team have brought to Asset Care and the HACS as a key element of that.

I was not involved in the adjudication of the awards but was very proud that Cambodia won the Golden Spanners award for HACS development, Figure 105.

Figure 105: Golden Spanners award received by Cambodia for HACS development for 2022.

GOVERNANCE AND PERFORMANCE: *What performance information do we need to tell us the status of our simplified and aligned processes and whether additional improvements are needed?*

Through the global Maximo system, we have as much information as we could need to compare the performance of the various breweries. In the APAC region I am part of a Steering Committee that meets quarterly and looks at the Planned Maintenance performance of all of the breweries in the region. From the Maximo system we know:

- Planned Maintenance Conformance to schedule: What percentage of Planned Maintenance schedules were completed according to the planned completion time.
- Breakdown Percentage: What is the total % lost time to breakdowns for the brewery.
- Equipment coverage: What percentage of equipment has the HACS in place.

- Integrated R&M cost/ARV: What is the cost of spares and labour as a percentage of Asset Replacement Value (ARV, explained in Chapter21).
- Planned Maintenance Work Orders closed: The number of workorders closed in the period.
- Spare parts Stock value: The total value of all spare parts in store.
- Planned Maintenance schedule backlog (hours): The number of hours of Planned Maintenance schedules outstanding.
- Stock value/ARV: The stock value as a percentage of Asset Replacement Value.

These are the performance indicators we have aligned on monitoring, but many more are available if needed. For example:

- Number of open service requests.
- Number of hours booked per month.
- Stock out ratio.
- MTBF (Mean time between failures).
- Line performance percentage.

In our reviews we look at these parameters for the region, for the breweries supported by the HACS Taskforce (A steering team that is supporting HACS development in several key breweries), and for each individual brewery.

INFORMATION AND SYSTEMS: *How do we need to change our systems so they can support our improved processes and process of improvement.*

In the past 3 or 4 years we have gradually implemented the latest version of Maximo in most breweries, and this version works through global networks with all performance data available to the Global Maintenance team.

Therefore, as described above, the local brewery and the Global Maintenance team are able to monitor performance and highlight areas of improvement, and this information feeds into the Governance and Performance category.

CHAPTER EIGHTEEN: SPARE PARTS MANAGEMENT

We need to have the right spare parts available at the right time in order to repair any unexpected breakdowns, to be able to carry out the necessary replacement activities as scheduled, and also to have the parts that we may need when an inspection activity reveals that the part must be replaced.

Building the HACS has allowed us to stabilize the requirement for spare parts, as many MSIs are now following a Time-Based Replacement strategy, meaning that we can predict when many parts will be needed.

One challenge that we face in Asia is that most of our equipment suppliers are in Europe, and Asian countries do not have the open borders that exist in Europe. When we import parts, different countries have different taxes and import duties, so it normally takes more than a week to clear parts through customs, even if they are flown in by air. By contrast, in Europe we can often obtain parts the next day, even from neighboring countries. The result of this is that we need to carry more spare parts stock in Asia than in Europe. It is also difficult to transfer parts from one brewery to another as in many countries an export license is required.

In our TPM process we have a Spare Parts Management route, the main points of this route follow, with material taken from the Spare Parts Management playbook developed by the Spare Parts Management Centre of Excellence.

18.1 ASSESS CURRENT SITUATION AND DEVELOP STRATEGY

First we need to assess where we are currently in terms of:

- Is the spare parts store safe to use and secure?
- Is there access control?
- Are there procedures for ordering, receiving and stock control of spare parts?
- Are there procedures for stock control of rebuilt or repaired parts?
- Are spare parts numbered according to Master data standards?
- Are the spare parts controlled in the CMMS?
- Is the CMMS system linked to the Procurement system (MRP) for automatic re-ordering of high turnover items?
- Is there a physical stock count of the inventory?
- Are obsolete items identified and written off?
- Are 5S Standards and visual management applied?
- Are the parts arranged logically?
- Do we know the value of parts and also the losses?
- Do we have climate-controlled storage for electronic and perishable parts?
- Do we manage the issuing of tools, lubricants etc.?
- How are consumables managed?
- Is there a work preparation area?
- Is there a spare parts receiving area?

We need to make sure that we have the required resources, and then we can develop a plan to close the gaps, reduce costs, and improve performance.

In the last few years, we have upgraded the condition of many of our Spare parts stores in Asia, and the process always has 2 main parts:

- Getting all the spare parts numbered and the data into the CMMS (Maximo) and linked to the MRP (Material Requirements Planning) system.

- Physically arranging the spare parts store.

18.2 IMPLEMENTING 5S IN SPARE PARTS STORE:

5S is the TPM methodology that we follow to organise and arrange the workplace: Sort, Set in Order, Shine, Standardise, Sustain. These steps involve examining everything in a workspace, deciding what's necessary and what isn't, putting things in order, cleaning, and setting up procedures for performing these tasks on a regular basis.

The SORT activity means throwing everything out that you do not need. One of the almost universal traits of Engineers that I have worked with is that they hate to throw anything away, in case they might need it one day.

Re-organising the spare parts stores requires a ruthless approach to removing all scrap, obsolete parts, broken parts and parts that one day might be needed but probably never will.

When I arrived in Cambodia in 2021 there were 90 containers of scrap materials left over from previous expansion projects. It has taken 2 years to reduce that to 8 containers of parts that we really (might) need. A useful trick is to obtain cans of spray paint, maybe a fluorescent colour, and jointly with the Engineers mark each item to be thrown out with a dot from the spray can. This is quick to do, and when you call in a scrap dealer to take away all the not-needed parts it is easy to identify and check what is to be thrown away.

Figure 106: Spare parts stores before re-organising, from HEINEKEN Spare Parts Manual.

Figure 106 shows some spare parts stores before 5S. Often TPM practitioners advocate to have a temporary storage area for the parts that are to be thrown out. This can be necessary if the parts have to be identified/confirmed by the Finance department and the financial write off has to be approved. This administration can be one of the challenges in reorganizing spare parts. Try to have good co-operation and support from Finance and try to avoid having a temporary storage area if at all possible.

Figure 107: Well organized Spare Parts Stores, from author's own collection.

Figure 107 shows 2 examples of very well organised spare parts stores.

Regular stock counts are important, and exactly how the parts are labelled and stored in the plastic bins can really facilitate this. For efficient stock count it must be easy to count the parts in each bin, such as by using dividers in the bin.

We have investigated some mobile applications that are able to look at a bin or stack of loose parts and count them, but so far we have not found these to be reliable.

Figure 108: Parts Identification boards.

Where we have many similar parts, such as for air connectors, bearings or wear-strips, it helps to have an identification board as shown in Figure 108. Using this board, a Technician who has removed a part from a machine can quickly identify the correct replacement.

Figure 109: Heavy parts in spare parts store, from author's own collection.

As shown in Figure 109, putting heavy motors and other heavy parts at floor level on moveable trolleys can greatly reduce the risk of injuries from manually handling the heavy parts.

Figure 110: Work Preparation area, from author's own collection.

A work preparation area, shown in Figure 110, is very useful to reduce the time that technicians spend looking for the required parts and tools for a particular job that is scheduled for a Planned Maintenance day. It is most useful for a planned Time-Based Replacement, as in the case of Inspection it is not known if parts need to be replaced or not.

For a Time-Based Replacement, all of the required parts can be put into a storage bin along with the Planned Maintenance card/Job Plan, along with any special tools, before the task is scheduled to be carried out. This reduces the time required to execute the task as the Technician does not waste time looking for the required parts or tools.

Figure 111: Control of special tools in our Thailand Brewery, from author's own collection.

Our brewery in Thailand has a very good control system for special tools kept in the store, such as lifting devices, torque wrenches or other tools not carried in the Technician's toolbox. This is shown in Figure 111. The tool card is taken out when the special tool is issued and moved to the "Borrowed Item" section. The tool card is replaced with a tag identifying the Technician who has borrowed the tool. When the tool is returned the Technician's tag is removed, and if the tool is not returned it can be easily seen which Technician has the tool.

A climate-controlled room is necessary for electrical components and perishable components such as rubber belts in Asia and Africa, as the high temperatures can lead to condensation forming on parts and they can quickly become useless. Normally an airconditioned room is sufficient to reduce the risk of condensation forming.

Access control is important, usually the store will be locked with only the storekeeper allowed inside to issue spares. This creates a problem at night, as we don't want to have 24-hour storekeeper coverage for occasional parts requirements, so there needs to be a system where Technicians can record the parts that they have taken after hours. However, Technicians are not always reliable at completing such administration, especially when there is pressure to repair a breakdown, so often we install cameras in the store so that if a Technician takes parts without completing the documentation he/she can be reminded.

In parallel with setting up the store physically, basic control processes and routines need to be set up, including:

- Coding all parts according to master data standards.
- Spare parts ordering process.

- Spare parts receiving process
- Stock count process and frequency.
- Return to stores process.
- Management of repairable parts.

18.3 RESTORE SPM BASIC CONDITIONS.

Once the spare parts store has been set up as described, we need to continuously work on whether we have the required spare parts in the right quantity.

It is common to classify spare parts according to demand frequency and price.

- High demand frequency, low price: Manage with automatic reordering and aim for high availability. Wholesale procurement model.
- High demand frequency, high price: Try to forecast demand and aim for moderate availability, LEAN procurement model.
- Low demand frequency, low price: Have the least impact, try to have lowest stock here, CLEAN procurement model.
- Low demand frequency, high price: Just in case or strategic parts. Here parts sharing/pooling with other breweries is logical, but we are not very effective at parts pooling yet.

Demand forecasting can be based on supplier information (but bear in mind that suppliers usually make more money from selling spare parts than from selling complete machines), or it can be based on historical demand.

Neither of these is going to be completely accurate because as we build the Planned Maintenance system we reduce unplanned downtime so there should then be less breakdowns and less spare parts needed.

One of the challenges I have come across is that in those breweries still following an annual overhaul approach to maintenance, the spare parts holding appears to be lower in the spare parts store control system because they purchase a large quantity of spare parts some months before the overhaul, as recommended by the supplier, and these parts do not go into the spare parts store stock (They are purchased against a specific Purchase Order and not part of regular MRP process). Sometimes this can be an amount of 500k USD or more for a packaging line or brewhouse overhaul. Some of the parts are used in the huge annual teardown exercise, and some are not and end up stored somewhere but not controlled as spare parts. In Cambodia we discovered over 2million USD of parts that had been purchased for overhauls and were not controlled by the stores stock control system (Maximo) (See Chapter 22).

You may ask why the left-over parts are not entered into the spare parts control system (Maximo), but the problem here is that Maximo is now linked to our financial control system, JDE. Adding these parts to Maximo manually would mean that we would then appear to have paid for the parts twice, once when we ordered them under the PO for the overhaul, and a second time when they entered the spare parts control system and the JDE system, which would double-count our actual expenditure. The alternative of entering the parts at zero cost is also not allowed under the financial rules.

When we move to Planned Maintenance instead of overhaul maintenance, all of the spare parts are properly accounted for in the spare parts management system, but paradoxically this can appear to our Financial colleagues to be an increase in spare parts stock holding!

To clarify further, If you overhaul 100 valves in an annual overhaul, you will buy 100 or more valve seal kits and change them all. None of them appeared in the spare parts management system (CMMS). When you follow Planned Maintenance, you will replace the valve seals according to a Time-Based Replacement schedule, and for these valves you might replace 2 seal kits a week, if they all have to replaced every year. With a 4-week lead time and a 2-week safety stock, you will now need to have more than 12 sets of valve seal kits in the store, whereas before you only had one or two to allow for unexpected failures.

This or similar examples occur in many breweries as we implement Planned Maintenance, and it can be hard to convince our accountants that we need to (apparently) increase spare parts stock levels, because of our previous bad habits of ordering masses of parts on separate Purchase Orders for annual overhauls.

The HACS helps significantly to improve demand forecasting for spare parts, as in each HACS COMPLETE file (the 5th Tab of the standard Excel template, see Section 21.1) we are able to see the parts that are planned to be replaced via a Time-Based Replacement strategy and when. If we then extract this from the CMMS we are able to forecast very accurately what is needed.

This does not cover parts that are subject to Inspection, and as these are likely to be higher value items we always have to take some guesses as to how much we need to keep in stock, depending on lead time.

We can use standard formulas to calculate re-order point and re-order quantity, which we build into our MRP system based on forecast consumption, lead time and re-order quantity.

All of the following steps in the Spare Parts Management improvement route are related to fine tuning the basics described above.

- Manage repairable items.
- Optimise stock levels and MRP settings.
- Manage slow moving parts.
- Implement spare parts pooling.

In Asia Pacific we have not yet realized savings from parts pooling, even though we have a great deal of SCALE equipment. For example, we now have 12 identical Gebo canning lines in the Asia Pacific region, but we have no parts pooling so every brewery has most of the parts that they might need, including expensive low frequency items, in their store. Part of the reason for this is the difficulty in moving spare parts from one brewery to another (In some countries a specific export license is required).

CHAPTER NINETEEN: PLANNED MAINTENANCE DAYS PLANNING

In order to carry out the INSPECTIONS, TIME BASED REPLACEMENTS, LUBRICATION AND CALIBRATION activities in the Planned Maintenance system (HACS), it is essential to have regular maintenance access to the assets, in our case packaging lines or the brewhouse, utilities plant or process equipment.

Fortunately, a lot of work had already been done by the Centre of Excellence (COE) for Planned Maintenance days, the following material is summarized from the COE playbook for Planned Maintenance Days.

To implement the HACS, we need a weekly Planned Maintenance day for each major asset group, i.e. Packaging line or Brewhouse.

- Planned Maintenance time should be 10% of the asset operational hours.
- The exact time required will depend on the Planned Maintenance schedules to be completed.
- Planned Maintenance for an asset should be the same day every week, and not on a weekend. Weekend maintenance is less productive as support staff are not available, local engineering companies may not be open etc.
- Planned Maintenance can only start when the area is cleaned and handed over by the production operators.

We still have some breweries in Asia Pacific where we do not have a regular Planned Maintenance day for each asset group (i.e.: packaging line), and maintenance takes place only when there is an opportunity. This occurs because of the misconception that stopping for maintenance will reduce production output, which is only correct in the short term. As we roll out the HACS we are convincing senior Managers that allowing time for regular Planned Maintenance means that we will reduce unplanned downtime and therefore total production will be increased.

If you refer back to the unplanned downtime table for Asia Pacific in Figure 25, you can see that unplanned downtime can be as high as 40% or more where maintenance is ineffective. We know from experience in Cambodia and other breweries that effective Planned Maintenance can bring unplanned downtime down to about 7% (with the included breakdowns down to below 5%). Therefore, investing 10% of production time in planned maintenance activities will still deliver a higher total production volume than not having Planned Maintenance. With effective Planned Maintenance we might have 7% unplanned downtime plus 10% planned maintenance time = 17%, which compares well with 30% to 50% unplanned downtime when there is no Planned Maintenance.

For the Planned Maintenance day to be effective, The Processes and Tasks in the HPO Planned Maintenance model must all be in place (see Chapter 17), including the Spare Parts Management (Chapter 18) and the Asset Care Standards (HACS).

The key to an effective Planned Maintenance day is **PPS**: Planning, Preparation and Scheduling. A good outcome will be that all of the planned work is completed on time and the asset returns to full operation without any losses of downtime or set up (we call this VSU, Vertical Start Up).

There are 3 steps to a well-executed Planned Maintenance Day:

19.1 PLANNING PREPARATION AND SCHEDULING

In the long term, there needs to be a plan of maintenance activities for the whole year. As explained in this book we need to minimize overhaul activity and maximise the execution of well-designed Planned Maintenance schedules, so there should no longer be any major overhaul activities taking place. There are a few machines that require a significant amount of dismantling to carry out some major Time-Based Replacement schedules, for example the seamer or the bottle washer in Packaging. Here a major intervention is needed, normally of a few days, but this should not be used as an excuse to plan a major overhaul at the same time.

If a major intervention is needed, we can use the downtime to complete other scheduled Planned Maintenance work, but not to carry out an unscheduled overhaul of other equipment opportunistically.

As soon as you plan a major overhaul in addition to the planned activities, you are going to bring in an unnecessary number of replacement parts with a high likelihood of infant mortality failure, and a high number of less experienced Technicians, and so there will be many human errors in the work as well.

The essential major interventions can be planned for the coming year, as can the regular Planned Maintenance days for each area. The plan should be shared with the production planning/S&OP team and aligned for the year.

Once you have an aligned Planned Maintenance plan for the year, we need to make sure that we have the resources to complete all of the Planned Maintenance schedules.

The HACS COMPLETE files (section 21.1) contain a worksheet that totals the maintenance hours required for the asset in each year and arranged by craft, shown in Figure 112. The required maintenance is also broken down across a 10-year cycle for the asset, because the maximum Time-Based Replacement period that we have is 10 years.

YC001 (Mechanic)

Frequency	Manhours
1M	0.83
2M	9.25
3M	6.08
6M	17.25
1Y	53.42
2Y	41.00
3Y	85.00
4Y	56.50
5Y	31.83
6Y	15.00
10Y	26.00

YC001 (Mechanic)

FREQ	Y1	Y2	Y3	Y4	Y5	Y6	Y7	Y8	Y9	Y10
1M	9.96	9.96	9.96	9.96	9.96	9.96	9.96	9.96	9.96	9.96
2M	55.50	55.50	55.50	55.50	55.50	55.50	55.50	55.50	55.50	55.50
3M	24.33	24.33	24.33	24.33	24.33	24.33	24.33	24.33	24.33	24.33
6M	34.50	34.50	34.50	34.50	34.50	34.50	34.50	34.50	34.50	34.50
1Y	53.42	53.42	53.42	53.42	53.42	53.42	53.42	53.42	53.42	53.42
2Y	-	41.00	-	41.00	-	41.00	-	41.00	-	41.00
3Y	-	-	85.00	-	-	85.00	-	-	85.00	-
4Y	-	-	-	56.50	-	-	-	56.50	-	-
5Y	-	-	-	-	31.83	-	-	-	-	31.83
6Y	-	-	-	-	-	15.00	-	-	-	-
10Y	-	-	-	-	-	-	-	-	-	28.00
TOTAL	177.71	219.71	262.71	276.21	209.54	318.71	177.71	275.21	262.71	279.64
Average Maintenance Hours/Week	3.70	4.56	5.47	5.73	4.37	6.64	3.70	5.73	5.47	5.90

Totalling of All Man Hours

FREQ	Weekly	Monthly	Annually
1Y	5.02	20.08	240.96
2Y	5.90	23.58	282.96
3Y	6.79	27.16	325.96
4Y	7.07	28.29	339.46
5Y	6.07	24.27	291.29
6Y	7.98	31.91	382.96
7Y	5.02	20.08	240.96
8Y	7.07	28.29	339.46
9Y	6.79	27.16	325.96
10Y	8.55	34.19	410.29
Average	6.63	26.50	318.03

YC002 (Automation Technician)

Frequency	Manhours
1M	-
2M	-
3M	-
6M	-
1Y	1.25
2Y	-
3Y	-
4Y	-
5Y	-
6Y	-
10Y	-

YC002 (Automation Technician)

FREQ	Y1	Y2	Y3	Y4	Y5	Y6	Y7	Y8	Y9	Y10
1M	-	-	-	-	-	-	-	-	-	-
2M	-	-	-	-	-	-	-	-	-	-
3M	-	-	-	-	-	-	-	-	-	-
6M	-	-	-	-	-	-	-	-	-	-
1Y	1.25	1.25	1.25	1.25	1.25	1.25	1.25	1.25	1.25	1.25
2Y	-	-	-	-	-	-	-	-	-	-
3Y	-	-	-	-	-	-	-	-	-	-
4Y	-	-	-	-	-	-	-	-	-	-
5Y	-	-	-	-	-	-	-	-	-	-
6Y	-	-	-	-	-	-	-	-	-	-
10Y	-	-	-	-	-	-	-	-	-	-
TOTAL	1.25	1.25	1.25	1.25	1.25	1.25	1.25	1.25	1.25	1.25
Average Maintenance Hours/Week	0.03	0.03	0.03	0.03	0.03	0.03	0.03	0.03	0.03	0.03

YC003 (Electrician)

Frequency	Manhours
1M	2.00
2M	-
3M	2.25
6M	-
1Y	29.00
2Y	1.00
3Y	-
4Y	-
5Y	16.50
6Y	-
10Y	48.00

YC003 (Electrician)

FREQ	Y1	Y2	Y3	Y4	Y5	Y6	Y7	Y8	Y9	Y10
1M	24.00	24.00	24.00	24.00	24.00	24.00	24.00	24.00	24.00	24.00
2M	-	-	-	-	-	-	-	-	-	-
3M	9.00	9.00	9.00	9.00	9.00	9.00	9.00	9.00	9.00	9.00
6M	-	-	-	-	-	-	-	-	-	-
1Y	29.00	29.00	29.00	29.00	29.00	29.00	29.00	29.00	29.00	29.00
2Y	-	1.00	-	1.00	-	1.00	-	1.00	-	1.00
3Y	-	-	-	-	-	-	-	-	-	-
4Y	-	-	-	-	-	-	-	-	-	-
5Y	-	-	-	-	16.50	-	-	-	-	16.50
6Y	-	-	-	-	-	-	-	-	-	-
10Y	-	-	-	-	-	-	-	-	-	49.00
TOTAL	62.00	63.00	62.00	63.00	60.50	63.00	62.00	63.00	62.00	130.50
Average Maintenance Hours/Week	1.29	1.31	1.29	1.31	1.68	1.31	1.29	1.31	1.29	2.72

Figure 112: HACS files maintenance hours summary worksheet, from HACS file

This is a very important feature of the HACS as it allows us to plan the Planned Maintenance activities for the year and make sure that we have the necessary resources, and it also allows us to know the manning required for the maintenance department.

For many years there has been extensive discussion about how many people there should be carrying out maintenance. From top management there is pressure to reduce costs and lower headcounts in the Engineering team, from the Engineering team frequently the view that more staff are needed to carry out the maintenance tasks. The discussions were always qualitative and somewhat emotional, as the Engineering Manager did not have a quantitative measure of the tasks and hours required.

Using the HACS files, we now know the man hours needed to carry out the Planned Maintenance schedules for the year, and we can size the maintenance team accordingly once the HACS is rolled out across the brewery.

However, in addition to the HACS execution time we also need to apply a 'hands on tools factor' to allow for AM support (supporting other TPM pillar improvements), BDA time, work preparation time. This factor could be another 10%.

In the medium term, i.e., the coming 3 to 6 months, we need to check that we have the spare parts and Technicians available for the planned activities, and we need to check with the production planning team that the Planned Maintenance windows that we have agreed are still going to be available.

In the short term we will plan the maintenance activities for the next week. The Planned Maintenance planner needs to confirm that the production plan for the following week has the agreed maintenance slots available.

Then the Planned Maintenance schedules are printed, and the tasks assigned to individuals. The assignment process greatly increases the task completion rate as the responsibility is clear. A board as shown in Figure 113 can be used to assign tasks.

Figure 113: Planned Maintenance work allocation board, from author's own collection.

The Planned Maintenance planner needs to check if any special safety permits are required (For example if working on a steam supply main).

For Time Based Replacement tasks, the parts and tools needed should be put together in the work preparation area to increase the execution efficiency (refer to section 18.2, Figure 110).

It should not be forgotten that there may be Corrective Maintenance Tags raised just before the maintenance day that would need to be addressed in the next maintenance day, so these must be planned for and prioritized as well.

19.2 MAINTENANCE DAY EXECUTION

Ideally at the start of a Planned Maintenance day there should be a short (15mins) kick off meeting attended by the Engineering team and the Production team.

In the meeting there should be alignment on:

- Timing of asset handover to Engineering and cleaning required before handover.
- Safety issues and LOTO reminder.
- Permits required if working from height or for hot work.
- Overview of activities planned for the day and schedule of activities.
- Spares availability.

During the maintenance day it is important to check that effective LOTO is in place at all times, the Engineering Manager, Team Leader and maintenance planner should be physically checking on progress and helping to resolve any issues.

Feedback on Planned Maintenance schedules is often difficult to obtain. Technicians infrequently put feedback comments on the Job plan, such as a modification to the steps in the Job Plan if they find them to be incorrect. They should be encouraged or even required to give feedback so that the Job Plan descriptions can be continuously improved and are not static. One approach we use in TPM is to offer a reward to the Technician who provides the most feedback each month.

As the HACS is rolling out globally we are getting feedback from breweries that find some issues as they implement the HACS, and we have a process in place to capture this and update the main HACS global database (Section 21.1).

Bloom advocates making RCM a living program (Bloom, 2006, p. 194) and proposes a formal system of feedback from the craftsmen/women on the quality of the Planned Maintenance schedules.

Another way to achieve this is to make sure that the Maintenance Planner is occasionally directly involved in Planned Maintenance schedule execution and is not just spending his/her entire time in a planning office, but that he/she is on the line checking the quality and execution of the Job Plans.

19.3 TESTING

This is the opportunity to eliminate a failure due to human error in the Planned Maintenance activity, and to achieve a Vertical Start Up, whereby the asset returns to full efficiency right after start-up.

Drive systems that have been inspected or replaced should be tested under full load.

Drive motors that have been replaced must be checked for correct direction of operation.

When conveyor wear strips or conveyor table-top chains are replaced, the effect of this must be considered. If there is an inspection device on the conveyor it will now need to be recalibrated as the relative height of the container on the conveyor has changed. If there is a transfer to another conveyor, that will need to be checked as the height difference may create a step leading to fallen containers.

Replacement of a Variable Speed Drive means that we must be careful to transfer the correct parameter settings from the previous VSD, and hopefully that was recorded somewhere securely.

Any change of PLC or similar control system means we must be sure to upload and re-install the correct software.

For conveyor settings the use of gauges is important. We have developed simple gauges to set the height of the container from the dating print head and to set the width of the conveyor guiderails. Use of such a gauge should be specified in the Job Plan.

One of the challenges in testing is that on a Packaging Line it takes several hours to start up the line and pass containers all the way through the line, and we have a similar problem in the Brewhouse. It is not reasonable to ask a Technician who has worked all day on Planned Maintenance schedules to stay several more hours and check the asset is working correctly, so we need to make sure that the job plans contain clear instructions on testing that the work has been correctly executed.

CHAPTER TWENTY: CONDITION MONITORING AND PREDICTIVE MAINTENANCE

Predictive Maintenance is currently receiving a lot of attention and there seem to be many Managers (and journalists) who believe that it is a quick way to reduce unplanned downtime without putting in the effort required to build a Planned Maintenance system.

Concepts like IOT and Industry 4.0 are creating expectations of a connected world where every device is able to tell us it's status and we no longer need so many maintenance activities.

Articles such as the one on the following link to the Wall Street Journal (Loten, 2022) claim that companies such as Augury can install wireless sensors throughout a production facility, upload the data to the cloud, analyse it with AI software and then relay real time insights to the plant maintenance team as to asset condition.

https://www.wsj.com/articles/predictive-maintenance-tech-is-taking-off-as-manufacturers-seek-more-efficiency-11662543000?li_fat_id=e804ce0b-fdf6-48d7-8dd7-b79ff0e80ed4

Whilst this is exciting, it raises a lot of questions, such as how to structure the coverage of the hundreds of sensors and know what they are listening to (So we still need a complete machine hierarchical structure down to MSI level), and if the sensor tells us a part is failing on a machine, we still need a CMMS that makes sure the part is in the store and we know how to change it. In addition, by adding so many sensors we have introduced many more parts that can fail and the sensors themselves have to be maintained.

So, I currently view the claims of an all-encompassing AI driven Predictive Maintenance system with some caution.

What we are currently able to do is to install Condition Monitoring on specific equipment items and then eliminate most or all of the Inspections and Time-Based Replacements that were assigned for that asset or MSI, trusting the Condition Monitoring to do that job for us, and we can link that Condition Monitoring into some form of AI to predict the life of the asset or MSI as well.

This has helped us to reduce manual maintenance activities to some extent, but it is an expensive process to do properly, and so far it has had to be built onto the basic framework of a well-structured asset hierarchy down to the MSI level, followed by an understanding of the likely failure modes of each MSI, as otherwise how do you know where to put the sensors, and what parts to replace when there is a problem ?

The IFM system described in the following section cost nearly 50 000 USD to provide Condition Monitoring on 8 Brewhouse pumps, so it is about 6000 USD per pump unit to install.

My fear is that organisations will take an unstructured approach to implementing Predictive Maintenance or Condition Monitoring, due to the excitement of such technology, and the sustained result will be similar to having unstructured maintenance: haphazard and not effective.

I advocate strongly that the implementation of Condition Monitoring and Predictive Maintenance must be done within a structured (hierarchical) framework so that we fully understand the asset, and the data that we collect on its performance.

In line with this, we have to also consider the progression of maintenance methodology:

Figure 114 from the HEINEKEN Global Maintenance team shows this progression of maintenance methodologies against the axes of Data and Reliability.

Before the development of the HACS, we had mostly sensory based inspections, an example would be a Job Plan that just says, "Check the motor", this is a sensory based inspection, as there is a completely subjective standard. The data obtained is limited and variable, and the consequent reliability is low.

PdM Journey

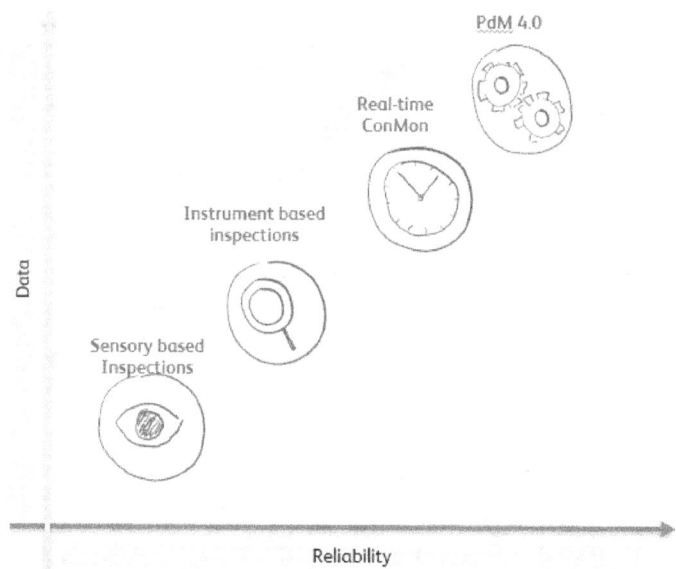

Figure 114: Predictive Maintenance Journey

With the HACS we have develop a library of instrument-based inspections. Every inspection has a standard and we use a range of tools as described in Chapter 15. This gives us more accurate data about equipment condition, and improved asset reliability.

The next progression is Condition Monitoring, which we have introduced on some of our equipment. In the following section I will explain our Condition Monitoring journey, and after that how to move to Predictive Maintenance.

But first we need to understand the difference between Condition Monitoring and Predictive Maintenance.

20.1 CONDITION MONITORING VS PREDICTIVE MAINTENANCE

Condition Monitoring is the process of monitoring one or more parameters in an asset's operation (vibration, temperature etc.), in order to identify any significant change which indicates that the equipment is in the potential failure part of the P-F curve. (Refer to Figure 26).

Predictive Maintenance is the process of collecting and analysing data (from Condition monitoring or elsewhere) in order to forecast the most cost-effective maintenance activities to maintain the condition of an asset and prevent breakdowns and may be conducted before a potential failure is detected.

Condition monitoring can be as simple as taking an oil sample from a gearbox. The oil can be analysed for pH, corrosive chemicals and metal particles, all of which will indicate the internal condition of the gearbox.

If the oil analysis tells us that there is a high metal content in the oil, then we can assume that the gears are worn, and we need to replace them. An engineer may look at the metal content in the gearbox based on this data he/she may schedule a replacement of the gears.

In the Predictive Maintenance definition above, key words are ANALYSE and FORECAST. A Predictive Maintenance system needs to be able to measure the condition, interpret the data against a standard and then take a decision to issue a replacement or other maintenance instruction.

Predictive Maintenance is therefore Condition Monitoring with some decision-making capability.

In the case of a gearbox, we might use a vibration sensor which can tell us when the gears or bearings are worn, and a Predictive Maintenance system would be able to analyse the trend in the vibrations and tell us when we need to overhaul the gearbox.

In their article "Why move from Condition Monitoring to Predictive Maintenance" (Mitchell, 2019), Yarmoluk and Mitchell give a good summary of the development of Condition Monitoring:

https://www.ibm.com/blogs/internet-of-things/iot-condition-monitoring-part-one/#:~:text=Both%20monitor%20the%20health%20and,or%2090%20days%20in%20advance.

The following is from their article:

CONDITION MONITORING 1.0 – 1980's AND EARLIER.

Legacy condition monitoring in an industrial environment has included lagging indicators like:

- Low lube oil pressure.
- High temperature.
- Irregular pump discharge pressure.
- Low or high seal pressure.

An alert condition on these measurements means a failure has, or is already, taking place and timely response is required. This is referred to as *"condition-based-reactive maintenance."* The indicator, although useful, does not give enough time to plan.

CONDITION MONITORING 2.0 – 1990's – 2000's.

A second wave of measurements have been adopted and have dramatically improved the detection of defects. Motor current, speed and power can be measured as a result of variable speed drives that have been deployed to improve efficiencies in electrical energy consumption. Additional vibration, bearing and temperature measurements are easier to achieve. The second wave of measurements include:

- Motor current.
- Speed.
- Power.
- Overall vibration.
- Bearing temperature.

A variance in any one of these measurements can indicate a condition that needs attention.

Using these measurements has proved fruitful for diagnosing problems. However, setting the alert thresholds for use with automated alerting has proved challenging. The nature of the process has made nuisance alarms common, varying process conditions require human analysis or some type of intelligence to identify a fault or anomaly in the normally varying measurements.

Overall vibration deployed in accordance with ISO alert standards has helped identify pre-existing conditions. Failure modes detected by overall vibration include:

- Imbalance.
- Misalignment.
- Looseness.
- Late stage bearing failure.

Overall vibration is a direct measurement for detecting and monitoring imbalance, misalignment and looseness of a rotating asset.

CONDITION MONITORING 3.0 – PREDICTIVE MAINTENANCE – 2010's

The monitoring described above could be good but is still lagging or condition based. For some, predictive maintenance is synonymous with technologies like:

- Infrared thermography (IR).
- Ultrasonic.
- Partial discharge testing.

This can be achieved by monthly vibration routes and analysis of vibration spectrum by trained and experienced professionals.

The failure modes targeted by this new intelligence includes 60- and 90-day advance detection of:

- Lubrication defects.
- Bearing defects.
- Cavitation.
- Pump seal failure.

In Condition Monitoring 3.0, overall vibration in combination with high frequency, or ultrasonic detection, provides the opportunity to realize true Predictive Maintenance. This is where a fault condition is identified 60 or 90 days in advance allowing operations and maintenance to plan and schedule a repair.

We are now entering Condition Monitoring 4.0. The capabilities presented here set the stage for the current state of condition monitoring and the introduction of artificial intelligence into the process. This is a dramatic transformation that has allowed innovative organizations to move from simple identification of a malfunction to proactive correction of the underlying problem.

The example given at the beginning of this chapter of the Augury product would be a Condition Monitoring 4.0 case.

20.2 CONDITION MONITORING IMPLEMENTATION

In our brewery in Cambodia, we started implementing Condition Monitoring in 2019, having selected IFM as the supplier of the instrumentation https://www.ifm.com/de/en

There are many companies offering a range of sensors for Condition Monitoring (and perhaps calling it Predictive Maintenance), but in selecting our supplier we had 2 main requirements that had to be met.

Firstly, that the data remained in our possession and was not in the cloud and then owned by the sensor supplier. It is a great business model for the supplier, to collect your data and then to charge you to access, analyse and interpret it, I recommend strongly that you retain full ownership of your data. It is the ownership of the data that has made the Amazon's and the Alibaba's of our world so successful.

The second requirement was that the sensors be hardwired and not Wi-Fi based sensors. Our breweries are large complex sites and in critical places like the pumping areas under the brewhouse or the compressor rooms in utilities there is a very high concentration of concrete, steel piping, pumps and heavy equipment that make a Wi-Fi signal very unreliable, so we decided on hard wired sensors that have a robust physical network to collect and process all of the data.

As they met the above requirements we chose IFM to install our Condition Monitoring system.

The next decision was where to start. In some breweries we have installed Condition Monitoring to solve a specific problem, for example to measure vibration on the bottle filler drive and monitor if a drive component needs to be replaced.

We did not want to follow a shotgun approach to implementation, and we wanted to implement Condition Monitoring in an area where the implementation cost would be repaid in savings, so we had to apply Condition Monitoring to fairly high value items. (We can put Condition Monitoring sensors all over the conveyors of the packaging line and monitor each one, but it will cost more than the value of the conveyors to do so).

In our brewhouse we have many pumps, some of which are small, but a few are of very large capacity and are not off-the-shelf items, i.e.: are not kept in our stores as spare parts. The failure of one of these pumps will cause a lot of downtime, and they are expensive, so we don't want to have spare ones in the spare parts store.

We decided to install the IFM Condition Monitoring system on the 8 pumps in the brewhouse that were the largest and were not off-the-shelf, these all had motors of over 15Kw.

As the experts on the sensing technology, IFM designed the sensor installation for each pump, one example is shown in Figure 115.

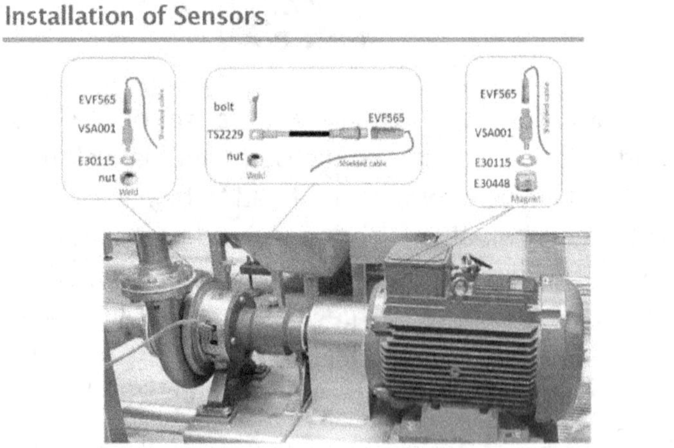

Figure 115: set up of sensors on brewhouse pump, from IFM documentation.

The MSIs of this pump include the bearings in the motor, the flexible coupling from the motor to the pump, the pump impeller and the mechanical seal in the pump.

The sensors are configured to monitor all of these MSIs, and they do this using vibration and temperature sensors.

IFM provide a Smart Observer monitoring package that allows us to view each pump's sensor output graphically as shown in Figure 116.

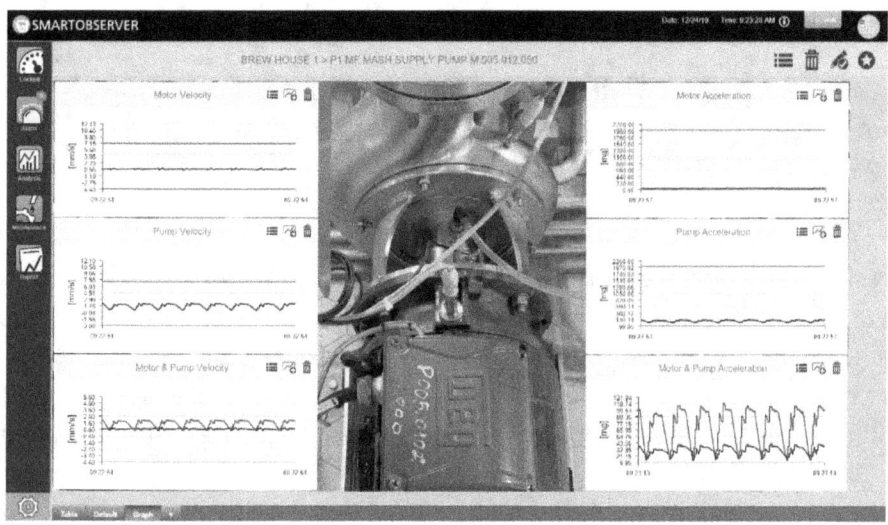

Figure 116: IFM SmartObserver monitoring package, author's own collection.

In the initial phase we set the alarm limits on each sensor according to ISO standard 13373 which gives approximate maximum values for vibration on rotating machinery.

However, on some pumps we see a high acceleration when the pump starts, especially if the liquid being pumped is viscous, and also vibrations when a tank runs empty, or a valve opens and closes in the piping system.

In Condition Monitoring it is our Technicians who look at the graphs and decide when action needs to be taken.

We have recently upgraded the monitoring system to run through our brewery central data collection system, and to be available as a dashboard in any of our control centers or in our daily department review meetings. In this application an algorithm has been added to the alarm parameters to allow for start-up or other incidental causes of vibration to be ignored.

The system works well and allows us to defer most inspections and Time-Based replacements on these pumps and to only take action when we have alarms from the monitoring system. We still have occasional false alarms, so we intend to further develop the algorithm to be more reliable before connecting the algorithm to the CMMS to automatically create a Planned Maintenance task when an alarm occurs.

The success and reliability of the system has been such that in 2021 we expanded it to 11 ammonia compressors and 4 CO_2 compressors, as shown in Figure 117.

Figure 117: Condition Monitoring sensor set up on cooling and CO2 compressors.

It should be noted that not all of the Planned Maintenance schedules on the above equipment are replaced by Condition Monitoring. It is still necessary to change air filters, check oil levels and change oil periodically, check and adjust drive belt tension and carry out other running adjustments that may not cause a Condition Monitoring alarm but are still required to keep the asset in good operating condition.

So, to repeat, a hierarchical structure to MSI level with a good analysis of likely failure modes and the maintenance actions required to prevent them is still needed.

Where Condition Monitoring has been most valuable in terms of investment payback is in allowing us to extend the recommended overhaul time of the CO_2 compressors. For example, the CO_2 compressor is supposed to be overhauled (New piston sleeves, rings, bearings etc..) after 5000 hours but with Condition Monitoring applied to the compressor we were able to extend this to 8000 hours operation, achieving a delay of the 75 000 USD overhaul cost to the next financial year (and a reduction in the overhaul cost per operating hour of 60%).

Of course, I don't like to have an "overhaul" at all, but for some equipment such as compressors there is a point where a Time-Based Replacement of all of the internal parts is needed, so it becomes an "overhaul" but it is not a total disassembly and teardown, it is a major intervention based on the Condition Monitoring data on vibration.

In our breweries with poor electrical supply the operation of gensets for long hours is very expensive and plagued with large sudden catastrophic failures. I myself have experienced many Genset failures in Africa, Myanmar, Papua New Guinea and even Cambodia and they almost always involve a connecting rod exploding through the crankcase and the total destruction of a machine that costs around 500k USD to replace. In the past I have implemented monitoring systems with frequent temperature checking of bearings and rocker box covers via Infra-Red guns, with limited success. The engines we use on our generators are required to run for 24 hours a day for months or years, and being reciprocating engines they eventually fail.

Given these challenges, we are starting a pilot to develop Condition Monitoring on our gensets in Cambodia this year and we will share this with other breweries as soon as it is successful.

Clifford Jones

CHAPTER TWENTY-ONE: IMPLEMENTING THE HACS

21.1 THE HACS COMPLETE FILE AND THE HACS DATABASE

A key role of the Steering Committee of the HACS project was in standardizing the HACS files produced by each working group, to ensure that they can be easily shared across the many breweries later on.

We designed an Excel file that we called the HACS COMPLETE file that has a standardised format and includes the following worksheets.

- Cover Sheet (Describes the asset and key specifications).
- How to (Describes how to use the HACS file).
- Structure Overview (The division of the asset into sections by GA or P&ID).
- Hierarchical structure (The detailed breakdown of the MSIs).
- Planned Maintenance schedules and Job Plans (The schedules and Job Plans in the format of the Maximo upload sheets).
- SOPs and Attachments.
- Maintenance hours required (Calculation of maintenance hours needed by month and craft to carry out all of the maintenance activities).
- Maximo code table.

A central database of all the HACS files was also constructed with support from the Global Maintenance team and this database is accessible worldwide in our company.

In the next section I describe how I personally evaluate a HACS file before it is uploaded into the database. The Global Maintenance team is documenting the assessment standards and expanding this capability.

21.2 CHECKING THE HACS COMPLETE FILE

I assess HACS files to a very high standard, because one file might be used at many breweries across the company. Also, the HACS is developed only once and should be used for many years to come, so it should be correct.

It is very important that the Hierarchical Structure is clear and logical, as this may be used by others to develop HACS for other machines. A clear hierarchical structure makes it easy to use the existing HACS file to develop the next one.

In designing the Planned Maintenance schedules, we assume 120hrs of equipment operation per week.

There are several conventions in the writing of the HACS standards:

- We only include the components/MSIs that we need to maintain. We do not include in the hierarchy parts that usually require no maintenance. For example, a pipe in the brewhouse is not included, as it needs no maintenance, but a pressure sensor on the pipe is included. Items that rarely fail like brackets, machine frame, guiderail supports are not included.
- We decided that manual ball valves fail so rarely that they can always have a Run To Failure Planned Maintenance schedule, but they are included as an MSI.
- We treat a motor and gearbox as one MSI called "motor and gearbox".
- In the Hierarchy we use drawings from the machine spare parts catalogue to help visualization.
- In the hierarchy the machine section is a single digit, the construction group is 2 digits (1.2 1.2.....) then assembly is 3 digits 1.1.2..... and so on.

- When we have identical sections, like pallet conveyor sections, we have a separate Planned Maintenance schedule for each section. In Maximo we link the Planned Maintenance schedules via a route so that they are all issued together at the same time. We NEVER have one Planned Maintenance schedule that refers to several sections, as this creates problems in maintaining the asset history.

The Structure Overview worksheet must show the machine broken down into the standard sections as shown in Figure 118. The Main Machine is always Group 1, infeed Group 2, outfeed Group 3, Motor Control Centre (MCC) Group 4 and Auxiliary equipment Group 5 (OEM devices such as an automatic lubrication system or a hot-melt glue system). For process equipment Infeed and Outfeed may be combined into a single Transfer System group.

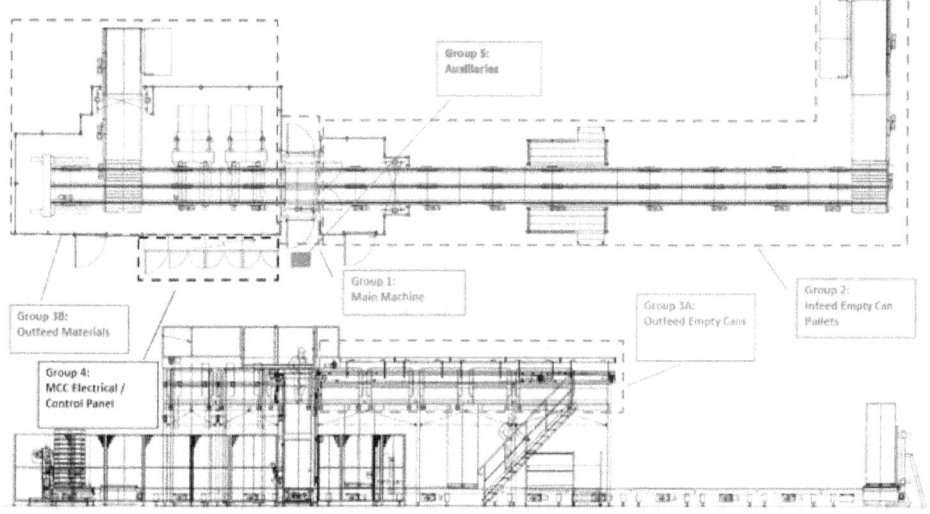

Figure 118: Overview of standardized machine sections in the HACS Structure Overview sheet.

Next I look at the Hierarchical structure sheet. The columns and format must be as shown in Figure 119. First, and very importantly, the Hierarchy must divide the machine sections into Construction groups (red arrows).

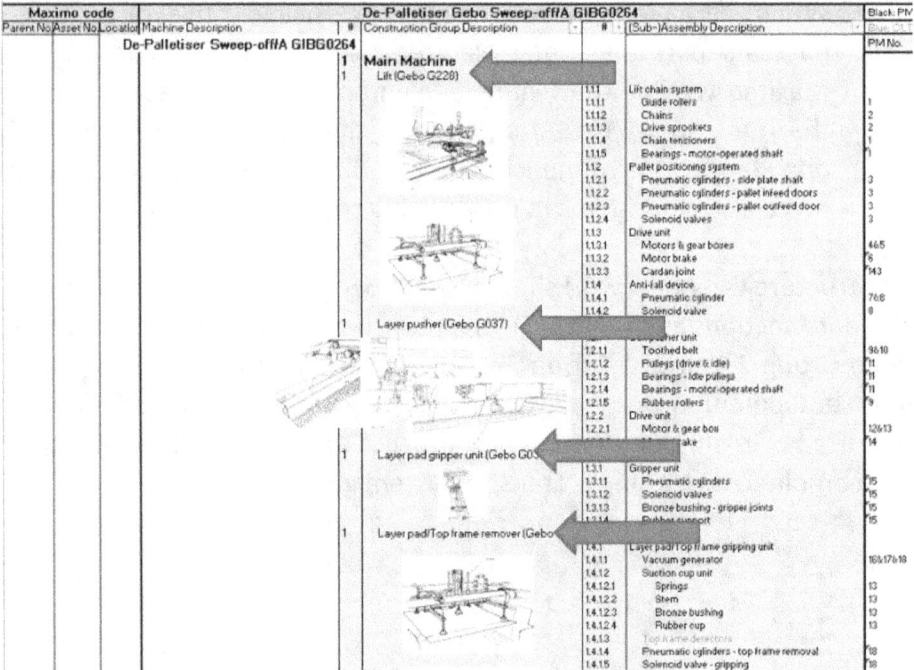

Figure 119: Hierarchical structure sheet in a HACS COMPLETE file.

Each Construction group must then be divided into assemblies and if necessary sub-assemblies:

In Figure 120 we can see the assemblies for the Lift Chain system, Pallet Positioning System, Drive unit and Anti-Fall device.

In the description we indent the sub-assembly, and we further indent the MSI.

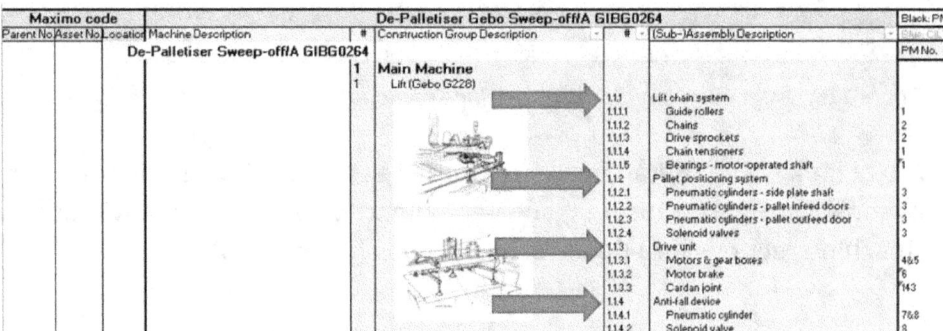

Figure 120: Hierarchical structure sheet in a HACS COMPLETE file.

Each assembly and if necessary sub assembly is further divided until we reach the MSI.

Every MSI must have a parent, grandparent and if necessary great grandparent that leads back to the main machine/asset: MSIs cannot suddenly appear in the structure, they must have a hierarchical path.

In Figure 121 the Assembly DRIVE UNIT is in the construction group LIFT, and there are 3 MSIs. Once we have identified the MSIs we can then develop the necessary Planned Maintenance schedule and Job Plan for each MSI. Here you can see there are two Planned Maintenance schedules for the Motor and Gearbox (Numbers 4 and 5).

De-Palletiser Gebo Sweep-off/A GIBG0264					Black: PM
#	Construction Group Description	#	(Sub-)Assembly Description		Blue, CILT
iIBG0264					PM No.
1	Main Machine				
1	Lift (Gebo G228)				
		1.1.1	Lift chain system		
		1.1.1.1	Guide rollers		1
		1.1.1.2	Chains		2
		1.1.1.3	Drive sprockets		2
		1.1.1.4	Chain tensioners		1
		1.1.1.5	Bearings - motor-operated shaft		1
		1.1.2	Pallet positioning system		
		1.1.2.1	Pneumatic cylinders - side plate shaft		3
		1.1.2.2	Pneumatic cylinders - pallet infeed doors		3
		1.1.2.3	Pneumatic cylinders - pallet outfeed door		3
		1.1.2.4	Solenoid valves		3
		1.1.3	Drive unit		
		1.1.3.1	Motors & gear boxes		4&5
		1.1.3.2	Motor brake		6
		1.1.3.3	Cardan joint		143
		1.1.4	Anti-rail device		
		1.1.4.1	Pneumatic cylinder		7&8
		1.1.4.2	Solenoid valve		8
1	Layer pusher (Gebo G037)				

Figure 121: Hierarchical structure sheet in a HACS COMPLETE file.

Electrical components on the machine itself, such as sensors are called FIELD ELECTRICAL.

They appear separately in the hierarchy at the end of each machine section.

This allows you to have a Planned Maintenance schedule to check photocells, for example, and all the photocells can be checked via those Planned Maintenance schedules that are issued as a group for that machine section, making the Technician's work easier.

Under Field Electrical we list/categorise first by DEVICE type, such as Photocell, and then list each photocell location (Figure 122).

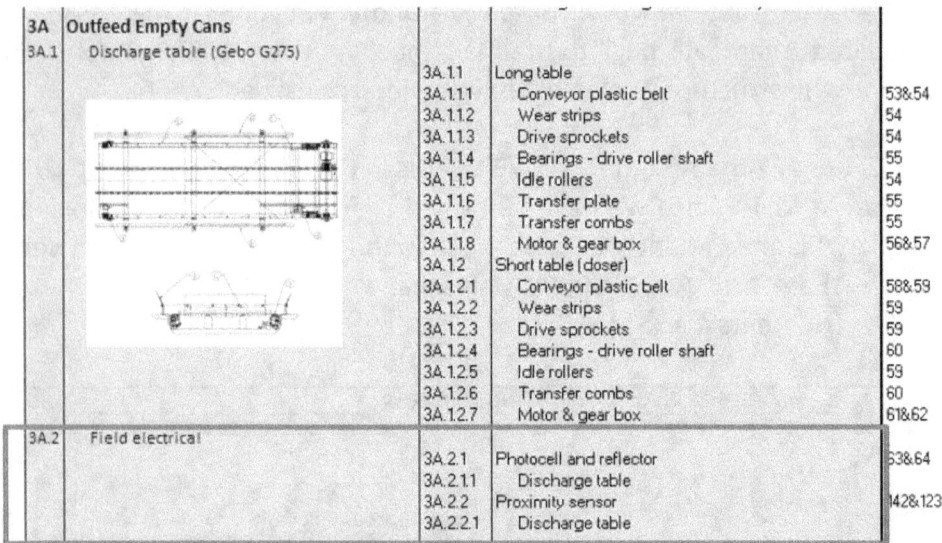

3A	Outfeed Empty Cans			
3A.1	Discharge table (Gebo G275)			
		3A.1.1	Long table	
		3A.1.1.1	Conveyor plastic belt	53&54
		3A.1.1.2	Wear strips	54
		3A.1.1.3	Drive sprockets	54
		3A.1.1.4	Bearings - drive roller shaft	55
		3A.1.1.5	Idle rollers	54
		3A.1.1.6	Transfer plate	55
		3A.1.1.7	Transfer combs	55
		3A.1.1.8	Motor & gear box	56&57
		3A.1.2	Short table (doser)	
		3A.1.2.1	Conveyor plastic belt	58&59
		3A.1.2.2	Wear strips	59
		3A.1.2.3	Drive sprockets	59
		3A.1.2.4	Bearings - drive roller shaft	60
		3A.1.2.5	Idle rollers	59
		3A.1.2.6	Transfer combs	60
		3A.1.2.7	Motor & gear box	61&62
3A.2	Field electrical			
		3A.2.1	Photocell and reflector	63&64
		3A.2.1.1	Discharge table	
		3A.2.2	Proximity sensor	142&123
		3A.2.2.1	Discharge table	

Figure 122 Hierarchical structure sheet in a HACS COMPLETE file showing Field Electrical.

To summarise, in evaluating the hierarchy I am looking for:

- No BOM items that are not MSIs (brackets, bolts etc.).
- All MSIs have at least one Planned Maintenance schedule (See Figure 123)
- Some MSIs will have more than one Planned Maintenance schedule, such as a pump, motor and gearbox or bearing (See Figure 123).
- MSI naming must be standardized as per the existing HACS files, no unexpected names.
- The hierarchy is a complete reflection of the entire machine (judged from the machine overview).
- The hierarchy is structured as described above, especially that construction groups are logical and contain all the assemblies of that group.

Next I look at the Planned Maintenance schedules and Job Plans:

EVERY MSI should have one or more Planned Maintenance schedules unless the strategy is Run to Failure or Condition Monitoring.

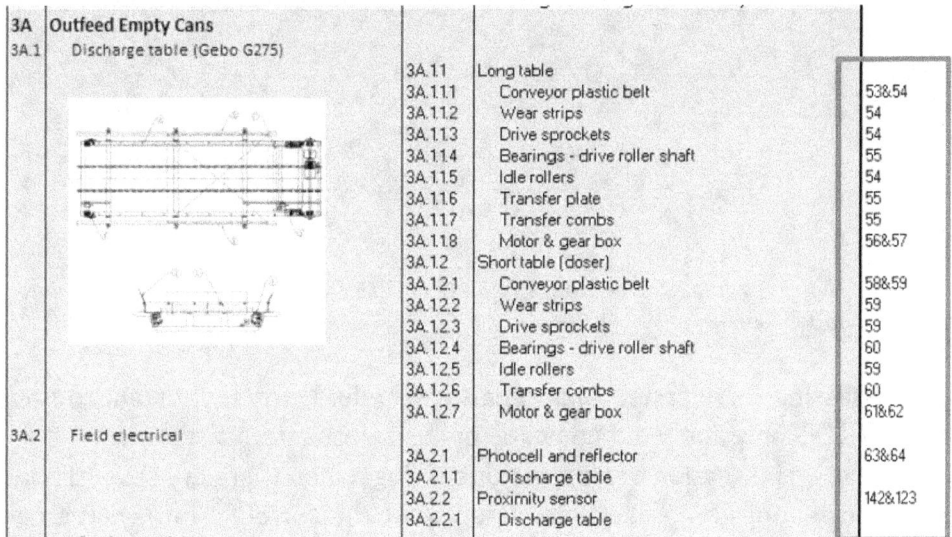

3A	Outfeed Empty Cans			
3A.1	Discharge table (Gebo G275)			
		3A.1.1	Long table	
		3A.1.1.1	Conveyor plastic belt	53&54
		3A.1.1.2	Wear strips	54
		3A.1.1.3	Drive sprockets	54
		3A.1.1.4	Bearings - drive roller shaft	55
		3A.1.1.5	Idle rollers	54
		3A.1.1.6	Transfer plate	55
		3A.1.1.7	Transfer combs	55
		3A.1.1.8	Motor & gear box	56&57
		3A.1.2	Short table (doser)	
		3A.1.2.1	Conveyor plastic belt	58&59
		3A.1.2.2	Wear strips	59
		3A.1.2.3	Drive sprockets	59
		3A.1.2.4	Bearings - drive roller shaft	60
		3A.1.2.5	Idle rollers	59
		3A.1.2.6	Transfer combs	60
		3A.1.2.7	Motor & gear box	61&62
3A.2	Field electrical			
		3A.2.1	Photocell and reflector	63&64
		3A.2.1.1	Discharge table	
		3A.2.2	Proximity sensor	142&123
		3A.2.2.1	Discharge table	

Figure 123: Hierarchical structure sheet in a HACS COMPLETE file showing Planned Maintenance Schedules.

In Figure 124 you can see a well written Planned Maintenance schedule:

The Planned Maintenance schedule has a reference number to the hierarchical structure (here 1.1.1) and a Planned Maintenance (PM) schedule number. (We expect to see about 100 Planned Maintenance schedules on a machine like a palletiser.)

The Planned Maintenance schedule description is standardised: Department, area, machine, construction group, assembly, frequency: PKG-CL-DEPALLETIZER-LIFT-LIFTCHAINSYSTEM-1Y

The Planned Maintenance schedule in Figure 124 is an inspection, carried out annually, it requires one person and takes half an hour. The Job Plan is listed clearly (Figure 125) and refers to several SOPs in Swipeguide to describe how to do the work. Finally, the spare parts are named, spare part numbers are given, and the quantity needed.

An Inspection is the correct maintenance strategy for a chain. The chain stretch can easily be measured with a chain stretch gauge, and if it is stretched more than 3% it must be replaced. So, we have a clear standard.

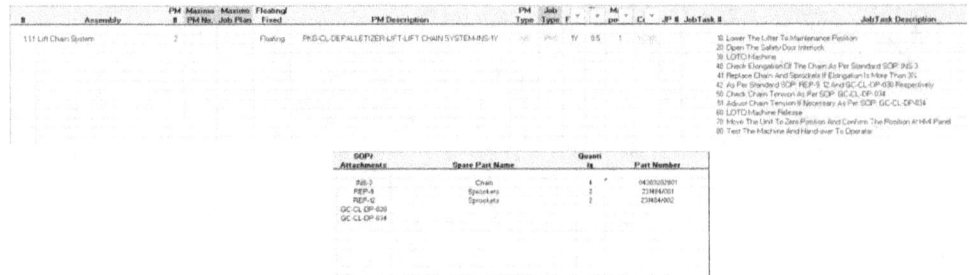

Figure 124: PM schedule sheet in a HACS COMPLETE file showing a Planned Maintenance schedule and Job Plan.

In assessing the Planned Maintenance schedule strategy, I want to see that INSPECTIONS only apply to high value or easily inspected parts.

To have an inspection of a conveyor bearing, when there are several hundred conveyor bearings on a packaging line, is not the correct strategy. The correct Planned Maintenance strategy for a conveyor bearing is a Time-Based Replacement. The frequency can be changed by the local brewery, but 5 years might be normal here.

I expect to see Time Based Replacement for low value items.

For electronic items we often cannot predict the failure, so we might have Time Based Replacement, or Run To Failure if the item is not important.

In the total Planned Maintenance schedules of a machine, not more than 5% of the items should be on a Run To Failure strategy.

Next I will look more closely at the Job Plan of the same Planned Maintenance schedule as shown in Figure 125.

The Job Plan steps are in increments of 10 so that additional steps can be added later.

LOTO is not the first step, because we have to lower the pallet hoist to the maintenance position, but after that there is a LOTO step.

Line 40 gives the main task, to check the elongation of the chain, and the detailed description is found in the SOP INS-3.

The standard here is a stretch of 3%. If more than 3% the instruction is to replace the chain AND the sprocket. Detailed description of how to do this is in the SOPs REP-9, 12 and GC-CL-DP-030.

SOP GC-CL-DP-034 tells you how to check the chain tension.

Finally, the machine is reset and handed back to the operator. This is a clear and well written Job Plan.

JobTask #	JobTask Description	JobTask Attachments
10	Lower The Lifter To Maintenance Position	INS-3
20	Open The Safety Door Interlock	REP-9
30	LOTO Machine	REP-12
40	Check Elongation Of The Chain As Per Standard SOP: INS-3	GC-CL-DP-030
41	Replace Chain And Sprockets If Elongation Is More Than 3%	GC-CL-DP-034
42	As Per Standard SOP: REP-9, 12 And GC-CL-DP-030 Respectively	
50	Check Chain Tension As Per SOP: GC-CL-DP-034	
51	Adjust Chain Tension If Necessary As Per SOP: GC-CL-DP-034	
60	LOTO Machine Release	
70	Move The Unit To Zero Position And Confirm The Position At HMI Panel	
80	Test The Machine And Hand-over To Operator	

Figure 125: Job Plan for chain inspection from a HACS COMPLETE file.

Examples of a badly written job plan are those that are not specific. For example, we often see:

- *Inspect the bearing.*
- *Inspect the motor.*

These instructions are not specific. It does not describe what you are inspecting for. We also see statements like:

- *Inspect the wear-strip for wear.*

Again, this is not specific, there is no standard. For a wear-strip we specify to check that the thickness is at least 50% of the original. Also, we have to instruct what to do if the inspection fails. Here is the correct statement:

Measure the thickness of the wear-strip at 3 equidistant points using a vernier caliper and if the thickness is less than 50% of the original, replace the wear-strip with a new one.

In summary, for Planned Maintenance schedules I am looking for:

- Each Planned Maintenance schedule is for a specific task and is not a teardown list for an overhaul.
- For a brewhouse vessel or a packaging machine I expect about 100 Planned Maintenance schedules. There can be up to 200 for the most complex machines, like a bottle washer or a multimodular labeller, and as low as 20 for a very simple machine like an ink jet coder or fill height detector.
- I will check to see that the Planned Maintenance schedules for OEM items like bearings, motors, pneumatic cylinders have been copied from the HACS database and not written from scratch. There are now

so many items in the database that there should be very few new/unique Planned Maintenance schedules, they should all be copied from earlier HACS files.

- I will check that not more than 5% of the maintenance strategy is Run to Failure.
- I will check that there is not an excess of Inspections compared to Time Based Replacement.
- Every inspection must have a clear standard. The standard should be measured with an instrument or gauge.
- The action to take if the inspection fails must be clearly stated.
- Every complex task should be linked to an SOP.
- LOTO must always be included.
- Job Plan numbering should be spaced in units of 10.

The part that requires experience is to assess whether the Planned Maintenance schedules are sufficient and appropriate to prevent the failure of the MSI. This comes with experience and knowledge of the equipment, but to help you there are some examples to refer to in Annexure Two.

Finally, I will check the SOPs. The SOP must show each important step of the maintenance activity, with sufficient detail that any qualified technician can carry out the task.

Figure 126 shows SOP INS-03, which explains how to use the chain stretch gauge to measure chain stretch. Each step has a photo and a description.

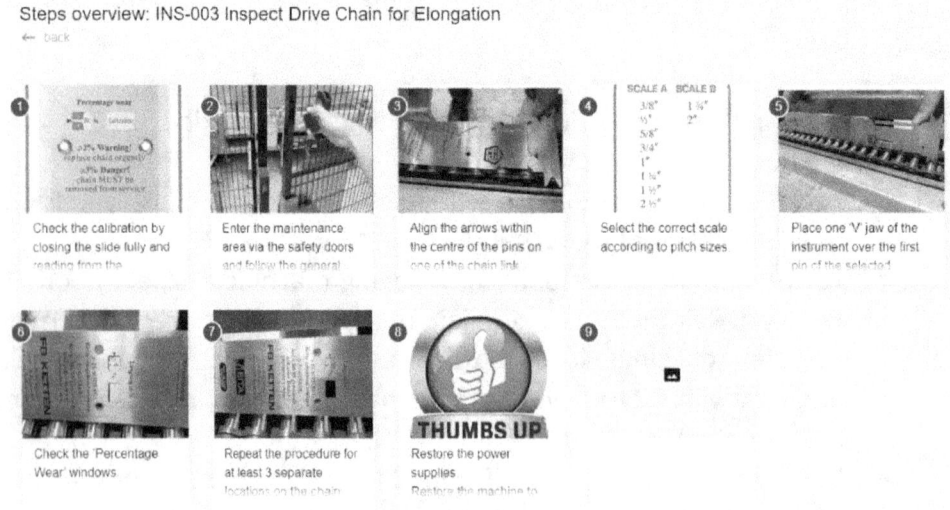

Figure 126: Example SOP for measuring chain stretch.

21.3 IMPLEMENTING THE HACS

Before implementing a Planned Maintenance system, you should have developed the required Planned Maintenance schedules for your assets (or even better copied them from another brewery), with all of the MSIs identified in a well-structured hierarchy and all of the necessary Planned Maintenance schedules should have clear Job Plans and SOPs to prevent the most common failures of those MSIs, and you should have also organised your Spare Parts Store and are having regular Planned Maintenance days.

When you activate the Planned Maintenance schedules you first need to consider the existing condition of your equipment.

If you have a set of assets such as a packaging line and it is nearly new or in very good condition then you can activate all of the Planned Maintenance schedules.

But if the assets are a few years old, you have to think about the activation. For example, perhaps a drive motor has a Time-Based Replacement of 5 years. If your asset is already 5 years old and you activate the Planned Maintenance schedules, it will be another 5 years until that drive motor is replaced. That means it will be running for 10 years and it is very likely to fail before replacement.

To avoid this problem, we carry out an RBC (Restore Basic Conditions) exercise on each asset group (packaging line, brewhouse) before we activate the HACS Planned Maintenance schedules. To Restore Basic Conditions we have to go through all of the MSIs and check whether they need to be inspected or replaced. This activity can end up being quite similar to an overhaul, but it is necessary to have everything in good condition before activating the HACS schedules, as they do not take into account any existing potential failure conditions: You need to reset all of the components back to the very beginning of the P-F curve.

(When an RBC is required, consider starting your Planned Maintenance Days with 'planned corrective' work, while developing your HACS schedules or preparing for implementation.)

The next consideration is the timing of the completion of the schedules. If you are going to introduce 1000 new Planned Maintenance schedules for a packaging line you don't want all of the 3 monthly inspections activating on the same day, so you need to plan the schedules in a way that spreads the workload evenly.

Figure 127 shows a small part of this planning, this is the HACS implementation table for one machine. This is then overlapped and staggered with all the other machines into a Masterfile to give the best smoothing of resources over the year. These files were developed by the Planned Maintenance Engineer at our Cambodia brewery.

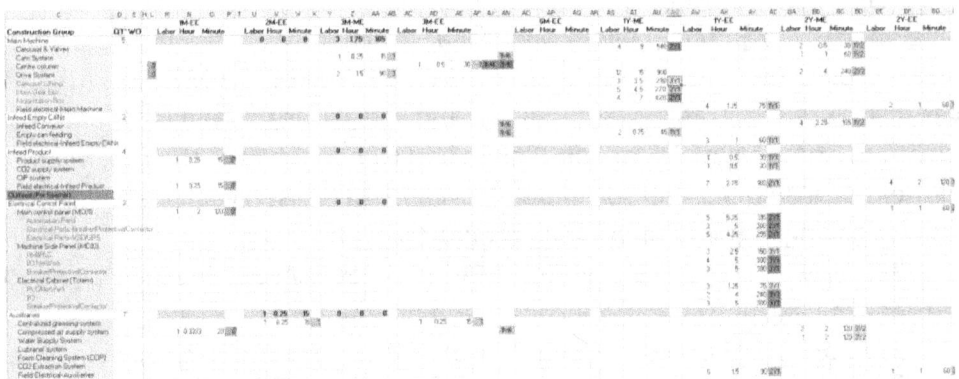

Figure 127: HACS implementation planning table.

A well-managed implementation is essential to build confidence in the new Planned Maintenance system. Issuing too many Planned Maintenance schedules in the first few weeks or months can create an impression of work overload for the Technicians.

Training of all Technicians is essential before the live implementation of the Planned Maintenance system. The training should include the basics of RCM, and some guidelines on completing the schedules, giving feedback and of course how to access the SOPs in Swipeguide. Following the training we usually have a launch event on the day of first implementation, a time to celebrate the new way of working and for the team to interact together.

Feedback from one brewery was that the Planned Maintenance schedule completion rate was higher when the HACS was introduced because the Planned Maintenance schedules were clearer, and it was easier for the Technician to understand what had to be done.

21.4 THE MAINTENANCE BUDGET

One frequent question I get from Engineers in our breweries is how to persuade senior management to allocate a bigger maintenance budget to the brewery.

As always, the answer is to have standards. For many years the standard in our company was to measure maintenance expenditure in Euro/Hl (Hl, Hectolitre = 100litres) produced. This meant that the largest breweries in Europe could achieve very low Euro/Hl maintenance costs, and this was often used as a comparison to the smaller breweries, who with less capacity had less economies of scale, and might be spending 3 or 4 times more per Hl on Maintenance. It was a benchmark that had no sound basis of comparison as there was such a great variance in the size, complexity and maturity of the 160 breweries across the company.

In 2020 HEINEKEN embarked on its Evergreen project to focus the organization on sustainable business priorities, and I was invited to lead a taskforce (including members of the Global Maintenance team) to design our strategy for Planned Maintenance, as there was no head of Global Maintenance at that time.

There was initially some pressure to reduce our Repair and Maintenance expenditure, but instead we focused on reducing our equipment replacement capex expenditure, assuming that if we could effectively implement Planned Maintenance, our assets would last longer and would not need to be replaced so often. I spent time thinking about aircraft like the DC3 or B52 that have flown more than 50 years and are still safe to do so, if Planned Maintenance is carried out correctly.

We knew that if we reduced the maintenance spend then the result might be more unplanned downtime. But if we implemented better Planned Maintenance (and in Asia I had the tool to do it, the HACS was by then progressing well), we could reduce the expenditure on replacement capex.

Our bold move was to reduce the expenditure on replacement capex by half (for mechanical equipment), promising that by implementing common sense Planned Maintenance we could double the lifespan of most of our assets.

However, not all replacement capex could be reduced as some is for electronic equipment that becomes obsolete, and we also had to allow time to phase in the implementation of the HACS Planned Maintenance system across the company. We came up with the target to reduce replacement capex by one third over 3 years, which is the principal delivery benefit of the HACS system, an amount of several tens of millions of USD.

At the same time, it was clear that we needed to standardize maintenance expenditure budgeting so that expenditure could be fairly compared across breweries. The industry standard for a world class organization is to spend between 2% and 3% of the Asset Replacement Value (ARV) on repair and maintenance (Mobley, 2008, p. 103).

We introduced the concept of measuring maintenance expenditure according to Asset Replacement Value, ARV, and this is now our global standard. The Asset Replacement Value is the cost of rebuilding the entire production facility in the case it is lost to fire or other disaster, and is the insurance valuation, provided by our insurers. For maintenance expenditure comparison we use the ARV of Plant and Equipment and exclude the value of the buildings.

This works well because the bigger capacity breweries will have a higher ARV, as will the ones with more complex equipment. The more there is to maintain, the higher the expenditure needed, but also the ARV is higher in a reasonably linear way. ARV is also a good measure because it is calculated and updated annually, by our third-party insurers, so there is no dispute over the correct ARV number for each brewery.

But not all breweries are at world class level, some are at the beginning of their improvement journey, so we allocated a higher percentage of ARV for maintenance expenditure to the less mature breweries, the percentage becoming lower as the breweries mature through the TPM process from Iron all the way to Gold level.

There is still some fine tuning to do in this system, such as to include an allowance for building maintenance.

By balancing across the global company and across regions, the allowed standards gave no net increase in Repair and Maintenance cost but achieved a more fair and defendable allocation of expenditure and sets targets for the breweries to achieve as they develop through the TPM process.

Shortly after this work was completed, Dennis van der Plas was appointed as the Head of Global Maintenance. Many Managers appointed to a new role like to revise all of the strategies developed before them, and to put their own stamp on them (usually renaming them). Dennis is sufficiently mature and self-confident to resist this practice, he has moved forward in his role supporting the HACS roll-out globally, as well as the strategy devised under Evergreen and the implementation of ARV without seeing the need to rebrand it. It is because of such maturity and openness that I asked him to contribute his foreword to this book, and I greatly appreciate his friendship and support.

21.5 HACS DEVELOPMENT PROGRESS

GEBO CANNING LINE

	PM's	SOP's
Palletiser	168	75
Depalletiser	147	50
Pasteuriser	90	17
Dome tray packer	117	46
Can Filler	132	49
Seamer	63	42
Conveyors	102	45
Can lids feeder	73	6
Date coders	20	14
Inspectors	62	20
Stretch wrapper	17	18
Others	29	8
	1020	390

MEURA BREWHOUSE

	PM's	SOP's
Materials treatme	371	
Mash Copper	184	
Mash Filter	178	
Holding vessel	79	
Wort Kettle	193	
Whirlpool	60	105
wort cooling	125	
CIP	116	
Dosing	202	
spent grains	33	
Others	362	
	1903	105

SCALE CELLARS

	PM's	SOP's
BMF Compact 2	343	
BMF 3	595	
Yeast Propagation	84	
Yeast storage/recovery	357	77
Fermentation	291	
Bright Beer	189	
CO2 recovery	223	
Stout cellar	261	
Schmidt Bretton Dealcol	313	
PVPP and Stout dosing	112	
DAW	134	
Alfa Laval Centrifuge	94	
GEA Centrifuge	44	
	3040	77

KRONES RGB LINE

	PM's	SOP's
Unpacker	91	
Washer	231	20
Krones EBI	133	47
Filler Crowner	193	43
Pasteuriser	45	44
Multimodular Lat	267	94
Packer	67	
Conveyors	140	18
FHI 1	23	2
FHI 2	38	3
	1228	271

KRONES CAN LINE

	PM's	SOP's
FHI 1 Myanmar	19	2
Can filler	144	30
Can seamer		
Can date coder	5	
Can lid feeder		
carton date cod	5	
conveyor	53	
Depalletiser	97	8
Full Case Inspe	23	
Full can inspec	23	
Packer	89	
Palletiser	99	
Pasteuriser	63	
Stretch wrapper	40	
	660	40

UTILITIES

	PM's	SOP's
Cooling	126	25
CO2 recovery	216	11
HVLV panels		
Air Compressors	28	
ERP water treatme	297	
conv water treaten	438	
WWTP	98	
	1203	36

KHS RGB LINE

	PM's	SOP's
Crate Depalletizer	63	
Unpacker	48	
Bottle Washer	371	
Crate Washer	49	
Filler & Crowner	99	
HEUFT EBI	63	
Pasteuriser	151	
Labeller - Krones	192	
Packer	58	
Crate Palletizer	78	
Conveyors	60	
Robot Crate Palletizer		
bOttle Depalletizer		
	1281	0

HEI FEI RGB LINE

	PM's	SOP's
Filler and Crowne	212	40
	212	40

MALT HANDLING

	PM's	SOP's
Container lifter	102	
Bucket elevator	105	
Magnetic seperator	30	
	237	0

OTHERS

	PM's	SOP's
Fire pumps	138	
	188	0

BREWING DATA BASE

424 Planned Maintenance schedules with job plans for all the common components found in brewing and process equipment

TOTAL PM'S	10872
TOTAL SOP's	959

Figure 128: HACS progress as of early 2024

Figure 128 shows the HACS and SOP development progress as of early 2024. The HACS Planned Maintenance system covers all of Brewing, Cellars, Packaging and Utilities. There are a total of nearly 11 000 individual Planned Maintenance schedules with nearly 1000 SOPs.

If I had known it would need to be this big at the beginning I do not think that I could have taken it on, considering that this is a part time activity in addition to my full-time job of Supply Chain Director.

If you are considering implementing a similar Planned Maintenance system it is a very significant undertaking. For our brewery in Cambodia that has a capacity of 6 million Hls per year with 2 Brewhouses and 4 large capacity Packaging Lines we expect to issue about 14000 Planned Maintenance schedules per year. To cope with this workload more than 80% of the Technicians work only on Planned Maintenance with only a few supporting the operations for any breakdowns or Corrective Maintenance work. This is of course the ideal situation, and every manufacturer should aim for this structure.

The HACS development was a result of a lot of hard work by many students from Singapore Institute of Technology, Trainees and Engineering colleagues, and I hope that we all learned something from the development process. Most of them are shown in Figure 129.

Of course, this large team did not work on the HACS full time, at any one time in the development phase there were perhaps 3 or 4 people working on the HACS for a 6-month assignment under direction of the Steering Committee.

Figure 129: Engineers and Students who contributed to the HACS development.

The HACS is by no means perfect, there are still SOP's to be added to some machines and other machines need to be revised and updated. As we roll out the HACS across the globe we receive feedback on items that need to be updated, so the database is continually being improved.

Luckily maintenance perfection is not needed in our industry, so the HACS is more than good enough to achieve a step-change in our Planned Maintenance capability, and to take us from the "annual overhaul" mentality of 50 years ago to a more common-sense approach to Planned Maintenance.

21.6 RCM AND HACS TRAINING

In 2019 the HACS was still an APAC regional initiative. I was able to implement the HACS schedules in Cambodia, where they had been developed, as the team there was convinced of the benefits. One or two other breweries (Singapore and Myanmar for example) were early adopters where the Engineers and Supply Chain Directors had the insight to realise that an RCM based Planned Maintenance program was going to reduce their unplanned downtime and maintenance costs.

But some others were not so interested or did not understand RCM. Some senior Managers do not believe that Planned Maintenance is a cost-effective approach and believe that only Corrective Maintenance is needed. Others think that Planned Maintenance will be magically replaced by Industry 4.0 and that every machine will have an array of intelligent sensors that tell us everything we need to know.

To win "Hearts and Minds" I developed a training course that I started to deliver to the breweries in the APAC region. The training course covers RCM theory, failure patterns, FMEA, and the structure and design of the HACS. It was usually delivered as a one- or two-day course and I completed 300 man days of training in 2019 at locations across the region. Gradually the course was improved and further developed as I learned more and incorporated feedback from delegates.

In 2020 I was requested by Singapore Institute of Technology (SIT) to lecture their Engineering students on RCM and Planned Maintenance and this course is now delivered annually.

My own university experience (MSc Brunel) and looking at the curriculum of SIT and other universities shows that university engineering syllabus's do not usually include very much, if anything, about RCM and reliability. The focus is much more on design, thermodynamics, fluid mechanics, project management, strengths of materials etc. RCM and Reliability is usually not even a single subject option within a Mechanical Engineering or Integrated Engineering degree.

Searching extensively, I cannot find any Universities offering bachelor's degrees or undergraduate courses in maintenance or reliability, though there are some master's courses available in Reliability (UCLA, Manchester University, Edinburgh University) but in less than half a dozen universities globally.

Dr Jun Jie Chong has realized the need for this at SIT/Newcastle University in Singapore and that is why we work together, and why I have such respect him as a professional educator and thus I asked him to write a foreword to this book.

I would love to work with more universities teaching young engineers the fundamentals of RCM and maintenance, I hope that this book can be one step to educate more Engineers in RCM and that they can build their insights into the machines and systems that they design.

In HEINEKEN the Global Maintenance team have produced a "playbook" guide on the HACS implementation and is updating the TPM Planned Maintenance Pillar route to take into account the HACS and how it should be developed and applied in our breweries.

Recently I delivered my RCM and HACS training course in Thailand at the request of the brewery, and the Global Maintenance Team took the opportunity to have the training course professionally filmed and edited.

This video has been developed into an E-learning training package (Figure 130) as part of our Supply Chain Academy that includes theory tests and assignments in addition to the video explanations and is being rolled out to our Engineers globally. Adam Spencer in the Global Maintenance Team has done a fantastic job of making my lectures into a professional training course and delivering the remote training globally.

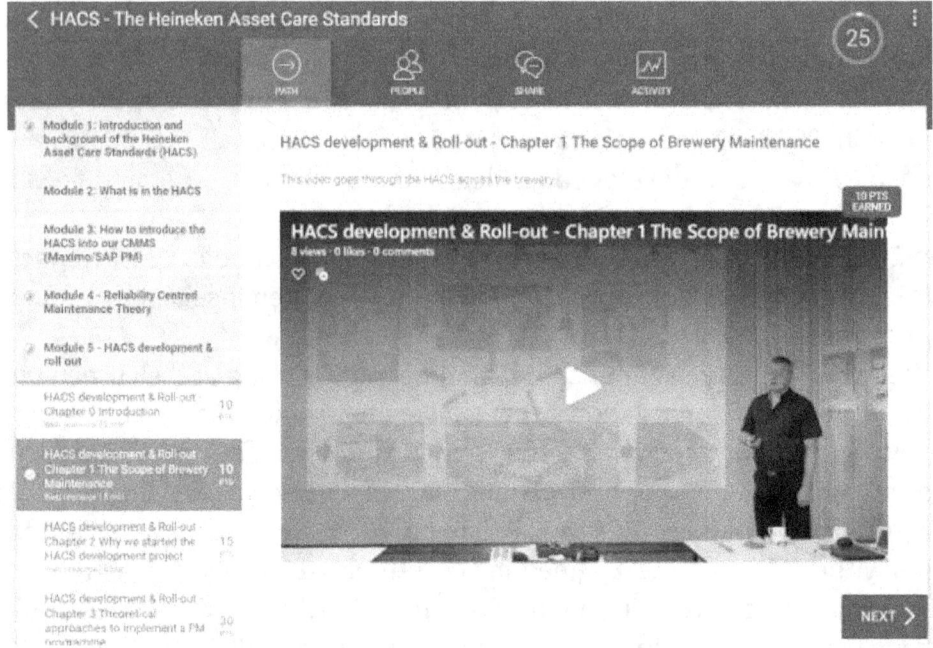

Figure 130: HACS E-learning course in the Supply Chain academy.

The training materials that I have developed and honed whilst building the HACS form the backbone of this book. I am very grateful to the senior managers in HEINEKEN who have authorized it's publication, believing as I do that we should share Common Sense Planned Maintenance with the wider beverage industry and also with Engineering students.

If you are interested in learning more about RCM or if you want to discuss collaborating to bring Reliability Engineering to your students, my e-mail addresses are jones.clifford@gmail.com and reliability-guru@outlook.com

CHAPTER TWENTY-TWO: KEY SUCCESS FACTORS AND BENEFITS OF THE HACS

There are some key factors that were in place in HEINEKEN without which the HACS could not have succeeded:

- CMMS systems were in place in every brewery, mostly Maximo.
- The Sahara project had put in place basics like spare parts stores and procurement procedures.
- SCALE EQUIPMENT: The SCALE project allowed the HACS to accelerate and roll out at high speed, especially at the beginning.
- We did not use any external consultants to design or develop the HACS.
- The project was fully supported by the Senior Regional Director, Jan Paul Boon.

There are many minor benefits and many qualitative benefits of the HACS system, such as having tools standardised, having SOPs available, having the same Planned Maintenance schedules for the same machines that are at different locations, most of them are explained in this book.

But for those who need to see the "hard "numbers, here are some of the quantified benefits:

1. Reduction in replacement capex by one third over 3 years across the organization (several tens of millions of dollars saving).
2. Reduction in annual Repair and Maintenance expenditure of 2 million Euros in our Cambodia brewery (Figure 131 shows a reduction of nearly 3 million Euro but some reduction was due to lower production in the Covid impacted years).

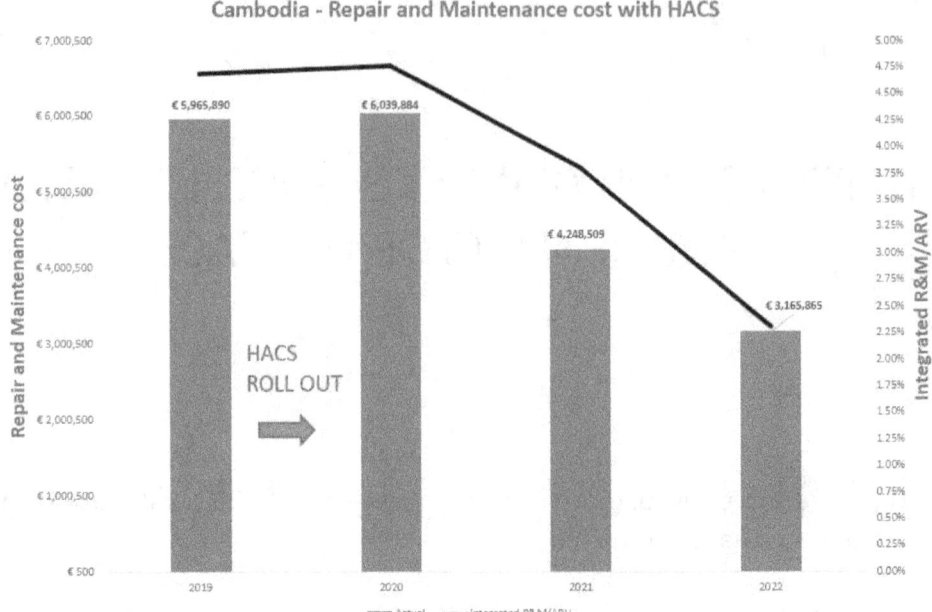

Figure 131: Repair and Maintenance cost reduction with HACS implementation from 2020 in Cambodia.

3. Recovery of 2 million USD of spare parts that was undocumented and left in the back of the stores as they had been ordered for previous overhauls and not used, so did not appear in our CMMS as spare parts. The team have now documented and built a separate store for these spares, shown in Figure 132, and they are now being used to further reduce our maintenance costs.

4. In 2021 we were contacted by one of our equipment suppliers and shown data that our spare parts purchasing from them had dropped drastically, and they wanted to know if we had stopped carrying out maintenance. I enjoyed explaining to them that this was because we had implemented an effective Planned Maintenance system.

Figure 132: Author (with mask) and Jan Paul Boon examine the 2mln USD of spare parts left over from previous overhauls and now integrated into our spare parts.

5. Unplanned downtime and breakdowns: Breweries with the HACS implemented are able to achieve breakdown rates across the brewery of 6% or less, and in this case the total unplanned downtime should be between 10 and 15% depending on other operational factors.

    Three breweries pioneering the roll out of the HACS in the APAC region are Cambodia, Singapore and Myanmar, their HACS implementation rate is shown in Figure 133. In Figure 134 their breakdown rate is shown.

    The operations are quite different, Singapore has a large number of SKU's and some aging equipment, Myanmar has some quite new equipment.

    We can conclude that the HACS can deliver breakdown rates of between 4% and 10%, but most importantly that the performance is sustained, and all 3 breweries show a gradual downward trend (January data point is missing for Cambodia).

    By contrast, in other breweries with no HACS we see breakdown rates of up to 25% .

This difference is not solely due to the HACS, we have to take an HPO approach as described in Chapter 17.

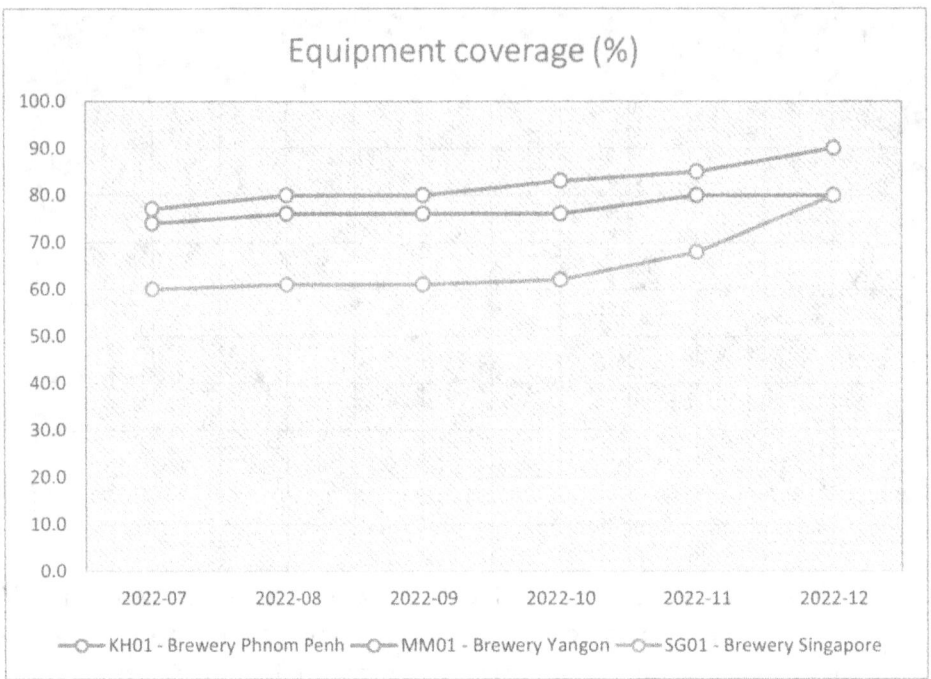

Figure 133: Equipment Coverage of the HACS in 3 APAC breweries, from Maximo dashboard.

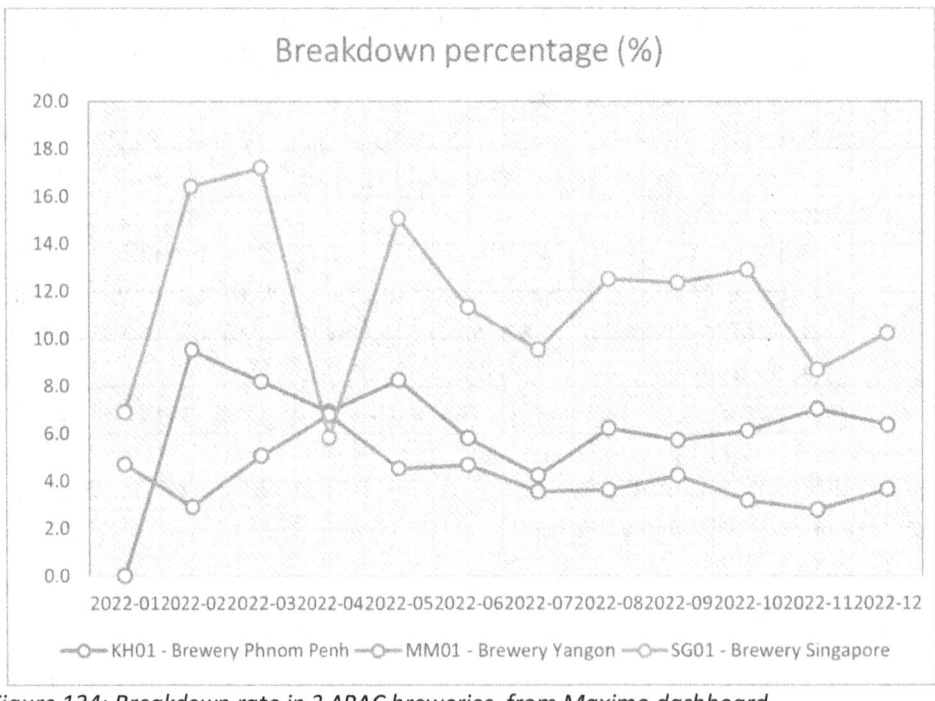

Figure 134: Breakdown rate in 3 APAC breweries, from Maximo dashboard.

ANNEXURE ONE: TERMS AND DEFINITIONS

DEFINITIONS

The TRADITIONAL ENGINEERING ASSUMPTION: *that every component has a certain lifetime after which it must be overhauled, inspected and if necessary replaced.*

A FUNCTION is the process, action or task that the asset, system or MAINTENANCE SIGNIFICANT ITEM is designed to perform.

A FUNCTIONAL FAILURE is when a MAINTENANCE SIGNIFICANT ITEM is unable to fulfill it's designed FUNCTION(s) causing the BREAKDOWN or reduced performance of a production asset.

A BREAKDOWN is when a production asset is unable to produce products in specification due to one or more failures of MAINTENANCE SIGNIFICANT ITEMS.

A FAILURE MODE is a description of the physical changes that caused the functional failure of a MAINTENANCE SIGNIFICANT ITEM.

A FAILURE EFFECT describes the impact that the MAINTENANCE SIGNIFICANT ITEM failure (failure mode) has on the assembly, system or asset of which it is a part.

A MAINTENANCE SIGNIFICANT ITEM is the lowest level component or sub assembly, that is likely to cause a BREAKDOWN of the asset in its useful lifetime, to which we apply a MAINTENANCE STRATEGY. It is either the part that is kept in the stores as a spare, or the lowest level of parts components available from the supplier.

MAINTENANCE STRATEGIES are the targeted maintenance activities carried out to prevent failure of a Maintenance Significant Item. They are Inspection, Time Based Replacement, Lubrication, Calibration, Condition Monitoring and Run To Failure.

A PLANNED MAINTENANCE SCHEDULE is a set of Planned Maintenance tasks that are carried out to prevent failures of Maintenance Significant Items. It consists of the Planned Maintenance Schedule that defines the task and frequency, and the Job Plan that describes the maintenance tasks step by step.

CONDITION MONITORING is the process of monitoring one or more parameters in an asset's operation (vibration, temperature etc.), in order to identify any significant change which indicates that the equipment is in the pending/potential failure part of the P-F curve.

PREDICTIVE MAINTENANCE is the process of collecting and analysing data (from Condition Monitoring or elsewhere) in order to forecast the most cost-effective maintenance activities to maintain the condition of an asset and prevent breakdowns and may be conducted before a potential failure is detected.

ABBREVIATIONS

5S: TPM methodology to organise and arrange the workplace: Sort, Set in Order, Shine, Standardize, Sustain.

5-WHY: An iterative interrogative technique used to explore the cause-and-effect relationships underlying a particular breakdown. The primary goal of the technique is to determine the root cause of a breakdown by repeating the question "Why?" up to five times. Also known as Breakdown Analysis (BDA) and Root Cause Failure Analysis (RCFA).

AD: Airworthiness Directives are legally enforceable regulations issued by the FAA to correct an unsafe condition of an aircraft.

AM: Autonomous Maintenance is an approach to equipment maintenance that involves giving machine operators responsibility for basic upkeep tasks.

APAC: Asia Pacific region.

ARV: Asset Replacement Value: The cost to replace all of the equipment assets in the brewery. Used as a standard for maintenance expenditure and usually updated annually by insurers.

BDA: Break-Down Analysis. See 5-Why

BOM: A Bill Of Materials is a list of the raw materials, sub-assemblies and parts, and the quantities of each needed to manufacture a product.

CAA: Civil Aviation Authority: The aviation authority in the UK that oversees aircraft airworthiness, the licensing of pilots, air traffic controllers, maintenance engineers, licensing of airports, and all other aviation standards.

CILT: Cleaning, Inspection, Lubrication and Tightening, the basic AM upkeep tasks carried out by machine Operators as part of the Preventive Maintenance program.

CM: Corrective Maintenance: Maintenance tasks caried out to correct an unplanned breakdown.

CMMS: A Computerized Maintenance Management System is software that centralizes maintenance information and facilitates the processes of maintenance operations.

CPU: Central Processing Unit: The part of a computer that interprets and carries out instructions in the software.

DoD: Department of Defence, a branch of the US government supervising all agencies of the United States Armed Forces.

FAA: Federal Aviation Authority: The aviation authority in the USA that oversees aircraft airworthiness, the licensing of pilots, air traffic controllers, maintenance engineers, licensing of airports, and all other aviation standards.

FMCG: Fast Moving Consumer Goods: products that sell quickly at relatively low cost, such as foods and drinks.

FMEA: Failure Modes and Effects Analysis is a step-by-step approach for identifying all possible failures in a design, a manufacturing or assembly process, or a product or service.

FTA: Fault Tree Analysis, a system wide evaluation to calculate the probability that an undesired event will occur.

GA: General Aviation: All Civil aviation excluding military aviation.

HACS: HEINEKEN Asset Care Standards, an end-to-end set of brewery Planned Maintenance schedules based on RCM philosophy.

HL: Hectolitre = 100 litres, standard unit of production or capacity in breweries

HPO: High Performing Organisation framework is a conceptual structure that managers can use for deciding what to focus on in order to improve organizational performance and make it sustainable.

I/O: Input/Output, the communication between a computer system and the outside world.

IC: An Integrated Circuit is a semiconductor wafer on which thousands or millions of tiny resistors, capacitors, diodes and transistors are fabricated.

ISO: International Standards Organisation.

LCD: Liquid Crystal Display, a type of flat panel display that uses liquid crystals for its operation.

KHS: German based supplier of Packaging machinery.

KRONES: German based supplier of Packaging and Brewing machinery.

LOTO: Lock Out Tag Out: The process of ensuring all forms of energy on a machine are isolated and secured with locks and notices to ensure the safety of maintenance staff or operators.

MCC: Motor Control Centre: An electrical panel containing the electrical controls required to operate motors, such as relays, isolators, starters and variable speed drives.

MRO: Maintenance, Repair and Operations.

MRP: Materials Requirements Planning, a system to calculate the materials and components needed to manufacture a product.

MSG: Maintenance Steering group, a taskforce formed by the American Air Transport Association, which consists of aircraft operators and manufacturers, to investigate Preventive Maintenance in the aircraft industry.

MSI: Maintenance Significant Item, see Terms and Definitions.

MTBF: Mean Time Between Failures.

MTBA: Mean Time between Assist.

NTSB: National Transportation Safety Board, an American Federal agency charged by Congress with investigating every civil aviation accident in the United States.

OEM: Original Equipment Manufacturer, a company producing components that are included in another manufacturers machine or equipment. Typically, sensors, motors, electrical components.

OPI: Operational Performance Indicator, a measure of operating efficiency.

OPL: One Point Lesson, a short, written instruction or procedure.

P-F: Pending Failure: The part of the failure curve where the pending failure can be detected with skilled inspection.

P&ID: A piping and instrumentation diagram is a detailed diagram which shows the piping and process equipment together with the instrumentation and control devices.

PCB: Printed Circuit Board, A pre-printed board used to connect electrical and electronic components together.

PCD: Pitch Circle Diameter, the diameter of the circle which passes through the centre of a set of points on the circle, such as the bolts on a wheel or the teeth on a sprocket.

PHE: Plate Heat Exchanger is a type of heat exchanger where product and cooling media are flowing in thin channels in alternate metal plates to maximise the effectiveness of the heat transfer.

PLC: Programmable Logic Controller, an industrial computer used for controlling processes in industry.

PM: Planned Maintenance, any maintenance activity that is planned, documented and scheduled.

PPS: Planning Preparation and Scheduling, particularly for Planned Maintenance days.

R&M: Repair and Maintenance: all maintenance activities including Corrective Maintenance and Planned Maintenance.

RBC: Restore Basic Conditions.

RCFA: See 5-Why.

RCM: Reliability Centered Maintenance, a concept of Planned Maintenance based on the 1978 publication of Nolan and Heap's Reliability Centered Maintenance.

RN: Royal Navy aka Her Majesty's Royal Navy.

RTF: Run To Failure, a maintenance strategy where no preventive maintenance action is taken as the MSI is not important to the operation. Often used when the failure pattern of the MSI is random.

S&OP: Sales and Operations Planning.

SAB: The South African Breweries, now part of AB Inbev.

SCALE: HEINEKEN project to standardize capital equipment purchases and reduce variation in the number of packaging line and brewhouse sizes that are purchased.

SIDEL/GEBO/CERMEX: An Italian manufacturer of packaging machinery.

SIT: Singapore Institute of Technology/Newcastle University in Singapore.

SOP: Standard Operating Procedure.

SWFP: Strategic WorkForce Planning, the process of identifying resources and competency requirements and developing a competency development plan to fill the competency gap.

TAG: A written label or notice, usually on a machine. It can be a safety notice to not start a machine, or also can be a request for maintenance activity.

TPM: Total Productive Management: An approach to reduce losses by dividing staff into functional Pillars and following a pillar route to attack all forms of losses at shop floor level.

TBR: Time Based Replacement, a maintenance strategy where an MSI is replaced based on a certain number of operating hours or cycles of the MSI. This implies that the failure pattern of the MSI is a wear-out curve.

UPD: Unplanned Downtime.

VSD: Variable Speed Drive.

VSU: Vertical Start Up.

ANNEXURE TWO: EXAMPLE PLANNED MAINTENANCE SCHEDULES AND JOB PLANS FOR FREQUENTLY OCURRING MSIs

The following are examples of the Planned Maintenance schedules and Job Plans for some of the most frequently occurring MSIs in a brewery:

CONVEYOR MOTOR AND GEARBOX

Conveyor drive motors are relatively low-cost small motors usually of 0.75kW capacity. The only Planned Maintenance schedules needed for a conveyor motor and gearbox are an oil change for the gearbox every 2 years (LUB) and a Time-Based Replacement (REP) with a complete overhauled unit, which is between 4yrs and 7yrs depending on the operational conditions.

Empty Can Mass Conveyor Drive Unit	3		Floating	PKG-CL*-EMPTY CAN MASS CONV DRIVE UNIT-LUB-2Y	LUB	PM2	2Y	0.25	1	YC001
Empty Can Mass Conveyor Drive Unit	4		Floating	PKG-CL*-EMPTY CAN MASS CONV DRIVE UNIT-REP-7Y	REP	PM4	7Y	2.00	2	YC001

Figure 135: Planned Maintenance schedule for motor and gearbox.

Figure 136 shows the respective job plans and SOP references. We rely on a CILT inspection by the operator to inform us of any oil leaks or visible problems.

10 Loto Machine
20 Change Motor Oil As Per LUB-1 and GC-CL-FI-022
30 Release Loto And Resume Program

10 Loto Machine
20 At the motor, note down which color wires are attached to which terminals
30 Replace the Drive Unit (Geared Motor) AS PER SOP: REP002
40 Fit the new Drive Unit and connect the wires as before
50 Loto Machine Release
60 Test run the conveyor to confirm correct directon of the motor

Figure 136: Job Plans for motor and gearbox.

In the case of a brake motor (a motor fitted with a mechanical brake) there is an additional Inspection (INS) of the brake unit to be carried out annually, the SOP in Figure 137 describes how to check the air gap on the brake disc.

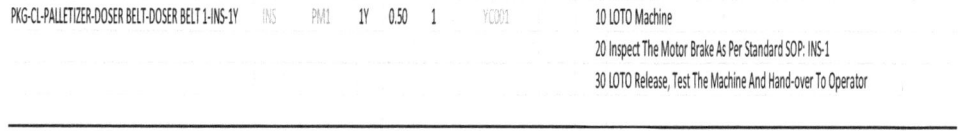

PKG-CL-PALLETIZER-DOSER BELT-DOSER BELT 1-INS-1Y	INS	PM1	1Y	0.50	1	YC001	10 LOTO Machine
							20 Inspect The Motor Brake As Per Standard SOP: INS-1
							30 LOTO Release, Test The Machine And Hand-over To Operator

Figure 137: Job Plan for motor brake.

CONVEYOR SLAT CHAIN

For plastic conveyor belt slat chain conveyors (canning lines) there is an annual Inspection (INS) of the chain and the supporting wear-strip, and an instruction for annual cleaning of the conveyor (CLE), as the dismantling and cleaning is important and complex.

Fixed	PKG-CL®-EMPTY CAN MASS CONV-INS-1Y	INS	PM1	1Y	0.25	1	YC001

Figure 138: Planned Maintenance schedule for mass conveyor.

The Job Plan and SOPs in Figure 139 instruct how to measure the stretch on the plastic chain (with a gauge) and the wear on the wear-strip, noting that if one wear-strip is replaced then they should all be replaced to prevent fallen containers on the conveyor.

10 Loto Machine	
20 Check The Elongation Of The Conveyor Plastic Belt As Per INS-6	INS-6
30 Check The Wearstrips As Per INS-8.	INS-8
31 If at least one wearstrips is worn >50%, the whole set of conveyor table wearstrips should be replaced.	
40 Release Loto And Resume Program	

Figure 139: Job Plan for mass conveyor.

This is combined with the yearly cleaning of the whole conveyor as per Figure 140:

10	Loto Machine
20	Remove the belt and clean it with pressure washer
30	Clean the frame and return ways with pressure washer
40	Inspect and Replace any damaged rollers, sprockets or wear strips
50	Install back the belt
60	Ensure that the slack on the belt is not below the conveyor frame
70	If the slack is below the frame, shorten the belt by removing links
80	Inspect all side guide rails for damage, bends, misalignment or wear.
90	Ensure side guide rails are all aligned/adjusted to allow for smooth can transfer.
100	Repair or replace side guide rails if necessary.
110	Set the guide rail distance for the correct nesting pattern using the special tool.
120	Place a steel ruler across the transfer between this belt and the downstream conveyor belt.
121	Check that the transfer is completely flat; if not, it may be necessary to inspect/replace the wearstrips.
130	Loto Machine Release

Figure 140: Annual cleaning for mass conveyor.

Because there are a large number of conveyor units on a packaging line, there should be separate Planned Maintenance schedules for each conveyor section, and the Planned Maintenance schedules should be staggered throughout the year. For example, one conveyor section may be maintained each week to balance the workload and not have a major overhaul of all conveyors once per year. If the annual overhaul approach is followed, it requires huge numbers of Technicians to do all that work in a few weeks, and there will be many errors leading to many infant mortality failures.

As the Planned Maintenance of the conveyors is quite standardized and repetitive, this is one set of maintenance tasks that can be successfully outsourced to a third-party maintenance organisation. In the (typical) case that Engineering resources are limited, outsourcing some routine and repetitive Planned Maintenance work to a trusted and qualified organization can be productive.

However, I don't recommend outsourcing more specific/complex Planned Maintenance tasks.

CONVEYOR SHAFT BEARING

Conveyor drive shaft and idler shaft bearings are lubricated (LUB) 3 monthly.

Fixed	PKG-CL*-EMPTY CAN MASS CONV-LUB-3M	LUB	PM2	3M	0.17	1	YC001	

Figure 141: Conveyor bearing Planned Maintenance schedule for lubrication.

The straightforward Job Plan for this is in Figure 142:

10	Loto machine
20	Remove the plastic cover of conveyor bearings on both sides of drive shaft and both sides of return
30	Clean the cover and bearings from any excess grease with a clean cloth
40	Grease the conveyor bearings: Apply 3 strokes of EP2 grease
50	Put the clean covers back on the bearings
60	Release the loto machine

Figure 142: Conveyor bearing Job Plan for lubrication.

There is a 5yr Time Based Replacement (REP) for the conveyor chains, sprockets, wear-strips, bearings etc..:

Infeed Plate Conveyor - Infeed Transfer System	113	REP	PM4	5Y	8	2	YC001	16	38

Figure 143: Conveyor bearing Planned Maintenance schedule for replacement.

Here we replace all of the chain sprockets, wear strips, idle rollers and shaft bearings as per the Job Plan and SOPs in Figure 144.

LOTO Machine	
Replace The Wore plate Conveyor Belt	
Replace The Sprockets	
Replace The Wear strips As Per SOP: REP-013	REP-013
Replace The Idle Rollers As Per SOP: REP-008	REP-008
Replace The Bearing As Per SOP: REP-007	REP-007
Adjust The Heigh Of Outfeed Conveyor As Per Manual Guide 6.3.2 (KR-BL-Pa-028)	KR-BL-Pa-028
LOTO Machine Release	
Test This Conveyor And Hand Over Production	

Figure 144: Conveyor bearing Job Plan for replacement.

PHOTOCELL AND REFLECTOR

For a photocell and reflector, we have an annual Inspection (INS), designed to check for damaged cables or loose connections.

Floating	PKG-CL-PALLETIZER-INFEED CARTONS-PHOTOCELL/REFLECTOR-INS-1Y	INS	PM1	1Y	0.50	1	YC003

Figure 145: Planned Maintenance schedule for photocell annual inspection.

In the SOP (INS-2) it is described to check the mounting bracket is not loose (a very common failure cause), clean the electrical connection and check for damaged wires, and that the reflector bracket is not loose (also a very common failure cause).

10	LOTO Machine	INS-2
20	Inspect The Photocell/Reflector As Per Standard SOP: INS-2	
30	Repeat For Other Photocells/Reflectors	
40	LOTO Release, Test The Machine And Hand-over To Operator	

Figure 146: Job Plan for photocell annual inspection.

We also have Time Based Replacement (REP) to replace the photocell every 10 years. This is because we cannot inspect for a pending failure of the device itself. As mentioned before, this replacement is really to update the device so can be extended to a longer period.

Floating	PKG-CL-PALLETIZER-INFEED CARTONS-PHOTOCELL/REFLECTOR-REP-10Y	REP	PM4	10Y	4.00	1	YC003

Figure 147: Planned Maintenance schedule for photocell replacement.

The straightforward replacement Job Plan is in Figure 148:

10	LOTO Machine
20	Power Shut-Off
30	Replace Photocells And Wire Connectors
31	Replace Reflectors
40	LOTO Release, Test The Machine And Hand-over To Operator

Figure 148: Job Plan for photocell replacement.

PT-100 TEMPERATURE PROBE

For the PT 100 temperature probe there is a Calibration Planned Maintenance schedule (CAL). In this case, on the pasteurizer the Calibration is annual, but it could be more frequent if the there is a higher consequence of failure.

| Floating | PKG-CL-PASTEURISER-TEMPERATURE SENSOR OF SPRAY WATER PIPES AND PUMP-CAL-1Y | CAL | FM3 | 1Y | 0.25 | 1 | CM4 |

Figure 149: Planned Maintenance schedule for PT-100 calibration.

The Job Plan in Figure 150 refers to the SOP which is also included in chapter 14.

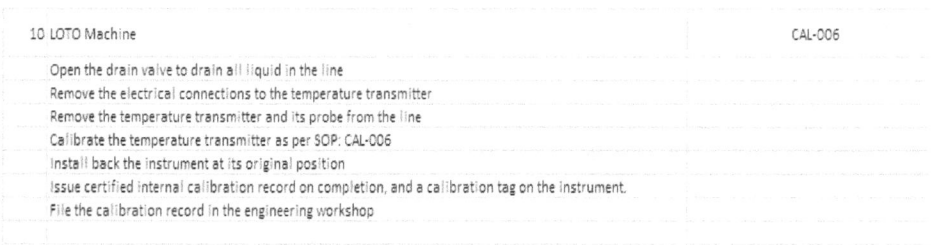

10 LOTO Machine   CAL-006

Open the drain valve to drain all liquid in the line
Remove the electrical connections to the temperature transmitter
Remove the temperature transmitter and its probe from the line
Calibrate the temperature transmitter as per SOP: CAL-006
Install back the instrument at its original position
Issue certified internal calibration record on completion, and a calibration tag on the instrument.
File the calibration record in the engineering workshop

Figure 150: Job Plan for PT-100 calibration.

There is no Time-Based Replacement of the temperature probe as the calibration will reveal when replacement is needed.

PUMP

The Planned Maintenance schedule for a (small) pump, in this case one of many on the pasteurizer, is a two-year Time-Based Replacement (REP) of the whole unit with an overhauled one. This is a fairly small size of pump, so Condition Monitoring is not applicable. Other pumps of the same size in less harsh environments have a 4-year TBR.

Motor & Pump -Pre-heating Zone (R1H)	9	REP	PM4	2Y	2	2	YC001	4	9

Figure 151: Planned Maintenance schedule for pump replacement.

The Job Plan in Figure 152 shows a straightforward procedure, with the link to the SOP which is a generic one.

10	LOTO Machine	
20	Replace The Motor &Pump 1 With Spare Motor&Pump From Store	
30	Replace The Motor &Pump 2 With Spare Motor&Pump From Store	
40	Replace The Motor &Pump 15 With Spare Motor&Pump From Store	
50	Replace The Motor &Pump 16 With Spare Motor&Pump From Store	
60	Overhaul & Replaced The Seals Of The Motor&Pump 1 & 2 & 15 & 16 At Workshop	
70	By Specialist As Per SOP: REP-027	REP-027
80	LOTO Machine Release	

Figure 152: Job Plan for pump replacement

Other larger pumps will have a different maintenance strategy that may include replacing bearings and seals and inspecting the impellers.

DOUBLE SEAT VALVE

For the double seat valves there are 3 Planned Maintenance schedules: Replacement of seals as a Time-Based Replacement (REP) every 5 years, Inspection (INS) of connections and operation every year, and Calibration of the valve operating controls every year (CAL). A typical brewery will have a few hundred of these valves, so like the conveyors the Planned Maintenance schedules should be staggered throughout the year, typically we will carry out the schedules on 3 or 4 valves every week to deliver the required level of maintenance.

Fixed	BRW-CB-Mature Beer Transfer System-REP-5Y	REP	PM4	5Y	1	2	YC001
Fixed	BRW-CB-Mature Beer Transfer System-INS-1Y	INS	PM1	1Y	0.5	1	YC003
Fixed	BRW-CB-Mature Beer Transfer System-CAL-1Y	CAL	PM3	1Y	0.25	1	YC005

Figure 153: Planned Maintenance schedules for double seat valve.

The Job Plans in Figure 154 give the steps for the Time-Based Replacement, the annual Inspection and the annual Calibration, with further details in the referenced SOPs:

10	LOTO Machine	REP-021, REP-22
20	Open The Drain Valve To Drain All Liquid In The Line	
30	Set The Valve Into Maintenance In Brewmaxx	
40	Remove The Double Seat Valve As Per SOP: REP-021	
50	Replace With A New/Overhauled Valve	
60	Overhaul The Old Valve Later In The Workshop To Replace The Actuator Seals As Per SOP: REP-022	
70	Set The Valve To Manual Mode In Brewmaxx	
80	Always Perform Autotune Function On Intellitop After The Replacement	
90	Fill The Production Line With Water	
100	If Leaking From The Flanges, Perform Retightening Until Leaking Stops.	
110	LOTO Machine Release	
10	Examine air pipes on the actuator for leaks using Ultrasonic Leak Detection as per SOP: INS-022	INS-022
20	If leaking, LOTO Machine and Compressed air supply and replace any leaking or damaged air pipes	
30	Set the valve into manual mode and check the opening and closing of the valve.	
40	If the valve does not open/close fully or takes more than 2 seconds to open/close.	
41	Remove the valve to check the seal and replace if necessary	
42	If seal is not damaged, replace the actuator	
50	Set auto mode for valve in brewmax	
10	LOTO Machine If Necessary	CAL-010
20	Perform Calibration Of Control Top As Per SOP: CAL-010	
30	LOTO Release	

Figure 154: Job Plans for double seat valve.

ACTUATED BUTTERFLY VALVE

For the automatic (actuated) butterfly valves there are 2 Planned Maintenance schedules: Inspection (INS) of connections and operation every year and replacement of seals as a Time-Based Replacement (REP) every 3 years, the frequency depending on the aggressiveness of the media in the valve.

If the media is not aggressive we would link the Time-Based Replacement to the Brewmax system and replace after 70 000 cycles of operation.

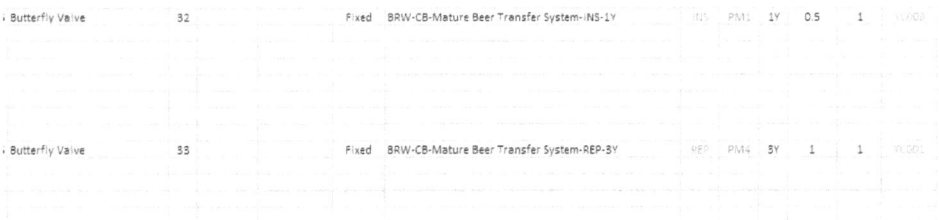

Butterfly Valve	32		Fixed	BRW-CB-Mature Beer Transfer System-INS-1Y	INS	PM1	1Y	0.5	1	W.X01
Butterfly Valve	33		Fixed	BRW-CB-Mature Beer Transfer System-REP-3Y	REP	PM4	3Y	1	1	W.X01

Figure 155: Planned Maintenance schedules for butterfly valve.

The Job Plans in Figure 156 describe the steps to be taken by the Technician for these two Planned Maintenance schedules with the links to the SOPs.

10	Examine air pipes on the actuator for leaks using Ultrasonic Leak Detection as per SOP: INS-022	INS-022
20	If leaking, LOTO Machine and Compressed air supply and replace any leaking or damaged air pipes	
30	Set the valve into manual mode and check the opening and closing of the valve.	
40	If the valve does not open/close fully or takes more than 2 seconds to open/close.	
41	Remove the valve to check the seal and replace if necessary	
42	If seal is not damaged, replace the actuator	
50	Set auto mode for valve in brewmax	
10	LOTO Machine	REP-016
20	Open the drain valve to drain all liquid in the line	
30	Set maintenance mode for the valve in Brewmaxx	
40	LOTO Compressed Air Supply	
50	Take note of the position indicator of the actuator if applicable	
60	Remove the electrical connections and mark them.	
70	Remove the old actuator from the valve body	
80	Replace with a new/overhauled actuator	
90	Overhaul the old actuator in the workshop as per SOP:REP-U16	
100	If applicable, ensure that the position indicator is pointing at the direction of the valve disc	
110	Install the Valve & Actuator into the line with its air and electrical connections	
120	Release Compressed air supply LOTO	
130	Use manual mode (in brewmaxx) to check the opening and closing of the valve.	
140	Close the drain valve	
150	Set auto mode for valve in brewmax	
160	LOTO Machine Release	

Figure 156: Job Plans for butterfly valve.

ANNEXURE THREE: GENERIC SOPs IN THE HACS SYSTEM

REP-001 Replace the Motor Brake
REP-002 Replace a Pneumatic Cylinder
REP-003 Replace an Anti-fall Device Pneumatic Cylinder
REP-004 Replace a Motor-Driven Pulley
REP-005 Replace a Linear Bearings/Ball Sleeves
REP-006 Replace a Plastic Conveyor Belt
REP-007 Replace a Transmission Bearing
REP-008 Replace a Pallet Conveyor Roller Chain
REP-009 Replace a Transmission Chain
REP-010 Replace a Pallet Conveyor Roller
REP-011 Replace a Solenoid Valve
REP-012 Replace a Drive Sprocket
REP-013 Replace Chain Wear Strips
REP-014 Replace Pallet Conveyor Chain
REP-015 Replace Butterfly Valve Seals (Pentair)
REP-016 Replace Actuator Seals for Butterfly Valve (Pentair)
REP-017 Replace Leakage Butterfly Valve Seals (Pentair)
REP-018 Replace Actuator Seals for Leakage Butterfly Valve (Pentair)
REP-019 Replace Butterfly Valve Seals (KSB Isoria 10)
REP-020 Replace Actuator Seals for Butterfly Valve (KSB Dynactair)
REP-021 Replace Double Seat Valve Seal (APV)
REP-022 Replacement of Double Seat Valve Actuator Seals (Pentair)
REP-023 Replace Gear Motor/Gearbox
REP-024 Replacement of motor without gearbox
REP-025 Replacement of Centrifugal Pump Motor and seal
REP-026 Replacement of chemical pump (Sigma2)
REP-027 Mechanical seal replacement of a Packo pump
REP-028 Replacement of the sealing packing
REP-029 Replacement of Modulating Valve Seals (Pentair S375)
REP-030 Replacement of Pneumatic Actuator SVP Seals (Pentair)
REP-031 Replacement of Vacuum Safety Valve Seals (Alfa Laval)
REP-032 Replacement of Seat Valve Seals (Schubert & Salzer)

REP-033 Replacement of Lamp Module on Turbidity Sensor (DTF16)

REP-034 Replacement of CO2 Sensor Seals

REP-035 Replacement of diaphragm and seal of Modulating Valve (CO2 Regulating Valve)

REP-036 Replacement of Seat Valve Seals (Sudmo)

REP-037 Replacement of Seat Valve Actuator Seals (Sudmo)

REP-038 Replacement of Pump Hose and Bearing (APEX)

REP-039 Replacement of O2 Transmitter Coating Holder (Haffmans)

REP-040 Replacement of Lamp Module (Optek)

REP-041 Replacement of Steam Trap Seals (Gestra MK-45)

REP-042 Replacement of Steam Globe Valve Seal (Conovalve)

REP-043 Replacement of Steam Diaphragm Valve Seal (Conovalve)

REP-044 Replacement of Steam Trap (Gestra UNA)

REP-045 Replacement of Agitator Shaft Bearing and Seal

REP-046 Replacement of Shaft Seals on Nord Motor Gearbox

REP-023B Replacement of Motors with Gearbox (Nord)

REP-047 Replacement of Butterfly Valve Seals (APV SVS/SV)

REP-048 Replacement of Pressure Exhaust Valve Seals

REP-049 Replacement of Mini Flow Valve (Actuated Needle Valve) seals

REP-050 Replacement of APV Double Seat Valve Actuator Seals

REP-051 Replacement of Higle HYGIA Pumps Mechanical Seal

REP-052 Replacement of Niezgodka Pressure Reducing Valve Seals

REP-053 Replacement of Pressure Relief Valve Seals (Kieselmann 6357)

REP-054 Replacement of Rotary Lobe Pump Gearbox Seals

REP-055 Replacement of Burkert Modulating Valve Diaphragm

REP-056 Replacement of BRAY 92 Actuator Seals

REP-057 Replacement of Higle SIPLA Pump Mechanical Seal

REP-058 Replacement of Sudmo S372 DSV Seals

REP-059 Replacement of Higle Durietta Pump Mechanical Seal

REP-060 Replacement of Pump Mechanical Seal (Alfa Laval Single Shaft Seal)

REP-061 Replacement of Pump Motor & Seal (Alfa Laval Single Shaft Seal)

REP-062 Replacement of Alfa Laval LKB Butterfly Valve Seals

REP-063 Replacement of Alfa Laval LKLA-T Actuator Seals

REP-064 Replacement of Alfa Laval SPC-2 Valve Seals

REP-065 Replacement of Kieselmann 6180 Vacuum Safety Valve Seals

REP-066 Replacement of Alfa Laval Mixproof Seat Valve Seals

REP-067 Replacement of APV Vacuum Safety Valve Seals (VRA11)

Clifford Jones

REP-068 Replacement of Handtmann Vacuum Safety Valve Seals
REP-069 Replacement of Alfa Laval Mixproof Seat Valve Actuator Seals

INS-001 Inspect Motor Brake
INS-002 Inspect a Photocell
INS-003 Inspect Drive Chain for Elongation
INS-004 Inspect Drive Chain for Tension
INS-005 Inspect Tension of Plastic Belt with Gravity Tensioning
INS-006 Inspect Conveyor Slat Chain (Stainless & Plastic) for Elongation
INS-007 Inspect Conveyor Plastic Belt with Rubber Pads
INS-008 Inspect Conveyor Belt Wear Strips
INS-009 Inspect Perimeter Guard Locks
INS-010 Inspect Transmission (Toothed) Belt Tension
INS-011 Transmission Toothed Belt Visual Check
INS-012 Inspect Proximity Sensors
INS-013 Inspect Wear Strips of Conveyor Chain/Roller Chain
INS-014 Inspect V-Belt Pulley for Wear
INS-015 Inspect V-belt for Tension.
INS-016 Inspect Conveyor Roller
INS-017 Inspect Tension of Plastic Belt with Tensioner
INS-018 Inspection and Cleaning of Silencer on Solenoid Valve
INS-019 Inspection of Steam Equipment using Ultrasonic Leak Detection
INS-019B Inspection of Steam Trap using Ultrasonic Gun
INS-020 Inspection and Cleaning of strainers
INS-021 Inspection of Diaphragm Rupture Sensors (Prominent Sigma Pumps)
INS-022 Inspection Using Ultrasonic Leak Detector
INS-023 Inspection of Pressure Reducing Valve (DP27)
INS-024 Inspection of Pneumatically Actuated Pressure Relief Valve
INS-025 Inspection of Pressure Relief Valves
INS-025B Inspection of Vacuum Safety Valves
INS-026 Inspection of Correct Motor Rotation
INS-027 Inspection of Anti-Vacuum Valve
INS-028 Inspection and cleaning of Switches with Plug-In Connectors
INS-029 Inspection of Level Switch (Negele NCS12)
INS-030 Inspection of Level Switch (E&H)
INS-031 Inspection of Flow Switch (IFM SI5000)
INS-032 Inspection of the packings tightening.

INS-033 Inspection of Level Switch (FTE30)
INS-034 Dial Gauge utilization
INS-035 Inspection of Flow Transmitter (Promag 10)
INS-036 Inspection of Flow Transmitter (Promag 50/53/Prowirl 72)
INS-038 Inspection of Flow Transmitter (Promass 40/80/83)
INS-039 Cleaning of Shell & Tube Heat Exchanger
INS-040 Inspection and Cleaning of Strainer
INS-041 Inspection of Water Filter Cartridge
INS-042 Inspection of Plate Heat Exchanger with Gappscan G2 Leak Detector
INS-043 Inspection of FTL50-1H Level Switch

LUB-001 Lubrication of Motor Gearbox
LUB-002 Lubrication of Chain
LUB-003 Lubrication of Centralized Lube System (Lincoln)
LUB-004 Lubrication of Roller Chain
LUB-005 Lubrication at Greasing Point
LUB-006 Oil Replacement of Chemical Pump Gearbox
LUB-007 Inspection of Gearbox Oil Level and Quality
LUB-008 Lubrication Using Ultrasonic Grease Gun
LUB-009 Oil Replacement of Nord Motor Gearbox
LUB-010 Replacing Lubricant in Bredel Peristaltic Pumps

CAL-002 Calibration of Conductivity Transmitter
CAL-006 Calibration of PT100 Temperature sensor
CAL-008 Calibration Pressure Sensor
CAL-009 Calibration Wort Density Measurement mPDS 1100
CAL-010 Calibration of Intellitop 2.0
CAL-011 Inspection of Pressure Gauge Accuracy
CAL-012 Calibration of Spent Grain Silos Level Transmitter
CAL-013 Calibration of a turbidity sensor Optek TF16
CAL-014 Calibration of Differential Pressure Switch (Filters)
CAL-015 Calibration of the silo weigher load cell
CAL-016 Calibration of the grist bin weight load cell
CAL-017 Calibration of Positioner (Burkert 8692)
CAL-018 Calibration of Modulating Valve Actuator (CO2 Regulating Valve)
CAL-019 Calibration of Gravity Transmitter (mPDS 5)
CAL-020 Calibration of Pressure Sensor (Haffmans)

CAL-021 Calibration of Oxygen Sensor (Haffmans)
CAL-022 Calibration of Turbidity Sensor (Optek)
CAL-023 Calibration of Turbidity Sensor (Haffmans)
CAL-024 Calibration of Burkert Positioner 8630
CAL-025 Calibration of Siemens PS2 Positioner
CAL-026 Calibration of Burkert Positioner 8694
CAL-027 Calibration of SAMSON 3730 Positioner
CAL-028 Calibration of ThinkTop (Basic)
CAL-029 Calibration of EaziCal IR Positioner (AVID)
CAL-030 Calibration of IP6100 Positioner (SMC)
CAL-030 Calibration of IndiTop (Basic)
CAL-031 Calibration of SIGRIST Phaseguard Turbidity Transmitter

ANNEXURE FOUR: SPECIFIC SOPs FOR THE SIDEL PALLETISER

GC-CL-PL MECHANICAL REPLACEMENT SOPs GEBO EVOFLEX GIFU0052

GC-CL-PL-001 Replacement of Toothed Belt at Motorized Layer Compactor

GC-CL-PL-002 Replacement of Servo Motor and Gearbox

GC-CL-PL-003 Replacement of Linkage Rollers at Motorized Layer Compactor

GC-CL-PL-004 Replacement of Motor and Gearbox at Split Roller

GC-CL-PL-005 Replacement of Pulleys and Bearings of Toothed Belt for Opening Roller
Carriage

GC-CL-PL-006 Replacement of Chains and Chain Guide Wheels for Opening Roller Carriage

GC-CL-PL-007 Replacement of Opening Roller Guides of Layer Deposit Module

GC-CL-PL-008 Replacement of Rollers for Opening Roller Carriage

GC-CL-PL-009 Replacement of Servo Motor and Gearbox at Layer Deposit Module

GC-CL-PL-010 Replacement of Transmission Belts for Layer Deposit Module

GC-CL-PL-011 Replacement of Tensioning Roller at Lifter Module

GC-CL-PL-012 Replacement of Toothed Drive Belts at Column Lifter Module

GC-CL-PL-013 Replacement of (main) Pulleys on Column Lifter Module

GC-CL-PL-014 Replacement of Retainer Rubber Sleeves of Gear Motor on Column Lifter
Module

GC-CL-PL-015 Replacement of Motors and Gearboxes on Column Lifter Module

GC-CL-PL-018 Replacement of Sprockets, Idle Rollers and Wear Strips on Short Metering Belt of Doser Belt

GC-CL-PL-016 Replacement of Linear Bearings of Both Arms on the Column Module 73

GC-CL-PL-017 Replacement of Gear Motors (Class 3)

GC-CL-PL-019 Replacement of Sprockets, Idle Rollers and Wear Strips on Long Metering Belt

of Doser Belt

GC-CL-PL-020 Replacement of Driven Support Rollers on Divider

GC-CL-PL-021 Replacement of Sprockets for Transmission Chain on Pallet Magazine (Arm)

GC-CL-PL-022 Replacement of Bearings (on Shaft) on Pallet Magazine (Arm)

GC-CL-PL-023 Replacement of Sprockets at Up/Down Drive Unit on Orthogonal 90Deg

Transfer Conveyor

GC-CL-PL-024 Replacement of Bronze Bushing at Up/Down Drive Unit on Orthogonal 90Deg

Transfer Conveyor

GC-CL-PL-042 Replacement of Toothed Belts for Front, Rear and Side Bars at Opening Roller

Carriage

GC-CL-PL-053 Replacement of Metering Belt Centering Roller at Doser Belt

GC-CL-PL-054 Replacement of Motor Driven Pulley of Split Roller Conveyor

GC-CL-PL-057 Replacement of Metering Belt at Doser Belt

GC-CL-PL-058 Replacement of Drive Unit Transmission Belt at Doser Belt

GC-CL-PL-060 Replacing the Bearings of Main Motor-Operated Shaft on Column Lifter

Module

GC-CL-PL-061 Replacement of Drive Sprockets at Forming/Accumulation Trolley

GC-CL-PL-062 Replacement of Conveyor End Roller at Forming/Accumulation Trolley

GC-CL-PL-063 Replacement of Metering Belt Drive Unit Bearings at Doser Belt

GC-CL-PL-065 Replacement of Ring Chains at Pallet Roller Conveyor

GC-CL-PL-066 Replacement of Bearings on Motor-Operated Shaft at Forming/Accumulation

Trolley

GC-CL-PL-067 Replacement of the Divider Sliding Blocks

GC-CL-PL-068 Replacement of the Divider Sliding Pipes

GC-CL-PL-069 Replacement of the Divider Chain and Chain Guide

GC-CL-PL-070 Replacing Transmission Chain of Up/Down Drive Unit of Orthogonal Transfer

GC-CL-PL-071 Replacement of Rubber Guide Rollers at Divider

GC-CL-PL-072 Replacement of the Guide Tracks at Divider

GC-CL-PL-073 Replacement of the Switching Guides at Divider

GC-CL-PL-074 Replacement of the Bearings (Fixed, Eccentric & for Arms) at Pallet Magazine

(Arm)

GC-CL-PL-075 Replacement of the Transmission Chain at Pallet Magazine (Arm)

GC-CL-PL-076 Replacement of Pallet Conveyor Rollers (Conveyor & Orthogonal Transfer)

GC-CL-PL-077 Replacement of the Split Roller Conveyor Drive Belt

GC-CL-PL-078 Replacement of Split Roller Conveyor Guide Rollers

GC-CL-PL-079 Replacement of the Split Roller Conveyor Tensioner Pulleys and Bearings

Inside

GC-CL-PL-080 Replacement of the Split Rollers

GC-CL-PL-082 Replacement of Roller Chain and Sprockets at Orthogonal Transfer

GC-CL-PL-084 Replacement of Roller Chain Wear Strips at Orthogonal Transfer

GC-CL-PL-085 Replacing the Layer Compactor Linear Slider Bearings

GC-CL-PL-086 Replacement of Chain and Wear Strips at Pallet Chain Conveyor and

Orthogonal Transfer

GC-CL-PL ELECTRICAL REPLACEMENT SOPs GEBO EVOFLEX GIFU0052

GC-CL-PL-025 Replacement of Push Button and Lamp (Reset Button) of HMI Operator Panel

GC-CL-PL-026 Replacement of HMI Screen

GC-CL-PL-027 Replacement of Air-conditioner/Ventilator

GC-CL-PL-028 Replacement of Input/Output Modules

GC-CL-PL-029 Replacement of Inverters/Variable Speed Drives

GC-CL-PL-030 Replacement of UPS

GC-CL-PL-031 Replacement of batteries in UPS

GC-CL-PL-032 Replacement of Power Supply Unit

GC-CL-PL-033 Replacement of SECROS 3 Motor Controller (Schneider)

GC-CL-PL MECHANICAL INSPECTION SOPs GEBO EVOFLEX GIFU0052
GC-CL-PL-034 Inspection of Roller Drive Belt Tension below Split Roller
GC-CL-PL-035 Inspection of the Coupling of Gearbox and Motor
GC-CL-PL-037 Inspection of Tension at Transmission Belt on Doser Belt
GC-CL-PL-038 Inspection of Rollers for Wear and Tear on Roller Conveyor
GC-CL-PL-039 Inspection of Tension of Drive Roller Chain (Long Chain) on Roller Conveyor
GC-CL-PL-040 Inspection of Rollers Movement on Pallet Roller Conveyor
GC-CL-PL-041 Inspection of Grease Distribution System for Blockage
GC-CL-PL-048 Inspection and Cleaning of Sliding Pipes and Blocks at Divider

GC-CL-PL ELECTRICAL INSPECTION SOPs GEBO EVOFLEX GIFU0052
GC-CL-PL-043 Functional Test for Safety (Photoelectric) Barriers
GC-CL-PL-044 Functional Test for Door Switches and Lightings for Control Panel

ANNEXURE FIVE: TECHNICIAN'S TOOLS LISTS

Mechanical Technician Tools List

N	Description	QTY
1	WRENCH, COMB, METRIC, 06mm	1
2	WRENCH, COMB, METRIC, 08mm	1
3	WRENCH, COMB, METRIC, 10mm	1
4	WRENCH, COMB, METRIC, 13mm	1
5	WRENCH, COMB, METRIC, 17mm	1
6	WRENCH, COMB, METRIC, 19mm	1
7	WRENCH, COMB, METRIC, 24mm	1
8	WRENCH, ADJUSTABLE, CHROME 250MM	1
9	SOCKET 3/8"DR, 6mm, 12PT	1
10	SOCKET 1/2"DR, 8mm, 12PT	1
11	SOCKET 1/2"DR, 10mm, 12PT	1
12	SOCKET 1/2"DR, 13mm, 12PT	1
13	SOCKET 1/2"DR, 17mm, 12PT	1
14	SOCKET 1/2"DR, 19mm, 12PT	1
15	SOCKET 1/2"DR, 24mm, 12PT	1
16	RATCHET, KNURLED HANDLE, 1/2"DR	1
17	EXTENTION BAR, 1/2"DR, 125MM/5"	1
18	RATCHET, ADAPTOR, 1/2FX3/8M	1
19	PLIERS, MULTIGRIP, 250MM/10"	1
20	PLIERS, COMBINATION, 213MM	1
21	PLIERS, LONG NOSE, 200MM/8"	1
22	PLIERS, CIRCLIP 180MM, INT STRT	1
23	PLIERS, CIRCLIP 180MM, INT BENT	1
24	PLIERS, CIRCLIP 180MM, EXT STRT	1
25	PLIERS, CIRCLIP 180MM, EXT BENT	1
26	PLIERS, LOCKING, CURVED JAW 7"	1
27	SCREWDRIVER, PHILIPS 3X150MM	1
28	SCREWDRIVER, PHILIPS 2X100MM	1
29	SCREWDRIVER, PHILIPS 2X35MM	1
30	SCREWDRIVER, BLADE 8X200MM	1
31	SCREWDRIVER, BLADE 6.5X100MM	1
32	SCREWDRIVER, BLADE 5.5X75MM	1
33	HEX KEY SET, MET 9PC	1
34	BALL PEIN HAMMER 350G/120Z	1
35	TIN SNIPS, AVIATION 250MM	1
36	TAPE MEASURE 10MX32MM	1
37	FEELER GAUGE SET METRIC, 0.05 to 1mm	1
38	CENTRE PUNCH 5MMX150	1
39	HACK SAW FRAME SQUARE 12"	1
40	UTILITY KNIFE 18MM SNAP-OFF	1
41	FILE 200MM SECOND CUT HALF ROUND	1
42	SCRIBER, ENGINEER DE 250MM	1
43	DIVIDER, SPRING, 200MM/8"	1
44	SQUARE, COMB. CARPENTER 300MM	1
45	FORHEAD MOUNTED LED LIGHT	1
46	KING TONY TOOL BAG, FABRIC 550x285x370	1

Electrical Technician Tools List

Nᵒ	Description	QTY	UOM	Craft
1	WRENCH, COMB, METRIC, 06mm	1	PC	MECH/ELECT
2	WRENCH, COMB, METRIC, 08mm	1	PC	MECH/ELECT
3	WRENCH, COMB, METRIC, 10mm	1	PC	MECH/ELECT
4	WRENCH, COMB, METRIC, 13mm	1	PC	MECH/ELECT
5	WRENCH, COMB, METRIC, 17mm	1	PC	MECH/ELECT
6	WRENCH, COMB, METRIC, 19mm	1	PC	MECH/ELECT
7	WRENCH, COMB, METRIC, 24mm	1	PC	MECH/ELECT
8	WRENCH, ADJUSTABLE, CHROME 250MM	1	PC	MECH/ELECT
9	SOCKET 3/8"DR, 6mm, 12PT	1	PC	MECH/ELECT
10	SOCKET 1/2"DR, 8mm, 12PT	1	PC	MECH/ELECT
11	SOCKET 1/2"DR, 10mm, 12PT	1	PC	MECH/ELECT
12	SOCKET 1/2"DR, 13mm, 12PT	1	PC	MECH/ELECT
13	SOCKET 1/2"DR, 17mm, 12PT	1	PC	MECH/ELECT
14	SOCKET 1/2"DR, 19mm, 12PT	1	PC	MECH/ELECT
15	SOCKET 1/2"DR, 24mm, 12PT	1	PC	MECH/ELECT
16	RATCHET, KNURLED HANDLE, 1/2"DR	1	PC	MECH/ELECT
17	EXTENTION BAR, 1/2"DR, 125MM/5"	1	PC	MECH/ELECT
18	RATCHET, ADAPTOR, 1/2FX3/8M	1	PC	MECH/ELECT
19	PLIERS, MULTIGRIP, 250MM/10"	1	PC	MECH/ELECT
20	PLIERS, LONG NOSE, 200MM/8"	1	PC	MECH/ELECT
21	PLIERS, CIRCLIP 180MM, INT STRT	1	PC	MECH/ELECT
22	PLIERS, CIRCLIP 180MM, INT BENT	1	PC	MECH/ELECT
23	PLIERS, CIRCLIP 180MM, EXT STRT	1	PC	MECH/ELECT
24	PLIERS, CIRCLIP 180MM, EXT BENT	1	PC	MECH/ELECT
25	PLIERS, LOCKING, CURVED JAW 7"	1	PC	MECH/ELECT
26	SCREWDRIVER, PHILIPS, INSULATED 1000V, 200MM	1	PC	ELECT
27	SCREWDRIVER, PHILIPS, INSULATED 1000V, 100MM	1	PC	ELECT
28	SCREWDRIVER, PHILIPS, INSULATED 1000V, 45MM	1	PC	ELECT
29	SCREWDRIVER, BLADE, INSULATED 1000V, 8X175MM	1	PC	ELECT
30	SCREWDRIVER, BLADE, INSULATED 1000V, 5X125MM	1	PC	ELECT
31	SCREWDRIVER, BLADE, INSULATED 1000V, 3X75MM	1	PC	ELECT
32	HEX KEY SET, MET 9PC	1	SET	MECH/ELECT
33	BALL PEIN HAMMER 350G/120Z	1	PC	MECH/ELECT
34	TAPE MEASURE 10MX32MM	1	PC	MECH/ELECT
35	CENTRE PUNCH 5MMX150	1	PC	MECH/ELECT
36	HACK SAW FRAME SQUARE 12"	1	PC	MECH/ELECT
37	UTILITY KNIFE 18MM SNAP-OFF	1	PC	MECH/ELECT
38	FILE 200MM SECOND CUT HALF ROUND	1	PC	MECH/ELECT
39	SQUARE, COMB. CARPENTER 300MM	1	PC	MECH/ELECT
40	PLIERS, 205mm, HIGH VOLT 1000V	1	PC	ELECT
41	CABLE STRIPPER 8-28MM	1	PC	ELECT
42	CRIMPER, PRE-INSULATED	1	PC	ELECT
43	SIDE CUTTER 200MM, 1000V	1	PC	ELECT
44	FORHEAD MOUNTED LED LIGHT	1	PC	MECH/ELECT
45	KING TONY TOOL BAG, FABRIC 550x285x370	1	PC	MECH/ELECT
46	FLUKE MULTIMETER	1	PC	ELECT

ANNEXURE SIX: REFERENCES

Basson, M. (2019). *RCM 3.* Industrial Press.

Bimetallic strip. (2012, fEB). Retrieved from Wikipedia: https://en.wikipedia.org/wiki/Bimetallic_strip

Bloom, N. B. (2006). *Reliability Centered Maintenance.* NcGraw Hill.

Chain Stretch Gauge. (2022). Retrieved from FB Ketten: https://www.fb-ketten.de/products/verschlei%C3%9Fmesslehre

Fault tree analysis. (2020). Retrieved from Wikepedia: https://en.wikipedia.org/wiki/Fault_tree_analysis

Field Instruments overview. (2023). Retrieved from Endress + Hauser: https://www.endress.com/en/field-instruments-overview

Frequency Gauge. (2023). Retrieved from Checkline: https://www.checkline.com/product/BTM-400PLUS

Gappscan. (2022). Retrieved from EIT International: https://www.eit-international.com/gappscan-flyer/

Grease Caddy. (2023). Retrieved from UE systems: https://www.uesystems.com/product/ultraprobe-401-digital-grease-caddy-pro/

Hanna, A. d. (2016, July). *hpo framework.* Retrieved from HPOcenter: https://www.hpocenter.com/article/hpo-model-hpo-framework-organizational-improvement-for-a-european-multinational/

Loten, A. (2022). *predictive maintenance.* Retrieved from Wall Street journal: https://www.wsj.com/articles/predictive-maintenance-tech-is-taking-off-as-manufacturers-seek-more-efficiency-11662543000?li_fat_id=e804ce0b-fdf6-48d7-8dd7-b79ff0e80ed4

Midler, P. (2010). *Poorly Made in China.* Wiley.

Mitchell, D. Y. (2019, March 15). *Blogs.* Retrieved from IBM.com: https://www.ibm.com/blogs/internet-of-things/iot-condition-monitoring-part-one/#:~:text=Both%20monitor%20the%20health%20and,or%2090%20days%20in%20advance.

Mobley, R. S. (2008). *Rules of Thumb for Maintenance and Reliability Engineers.* Elsevier.

Moubray, J. (1991). *RCM 2.* Butterworth Heinemann.

Nowlan, F. (1978). *Reliability Centered Maintenance.* San Francisco: Dolby Access Press.

Parsable. (2023). Retrieved from Parsable: https://parsable.com

Regan, N. (2012). *The RCM Solution.* New York: Industrial Press Inc.

Sealmaster. (2022). Retrieved from Rexnord: https://www.regalrexnord.com/brands/Sealmaster

Smith, A. M. (1993). *Reliability Centered Maintenance.* McGraw hill.

Sonic industrial imager. (2023). Retrieved from Fluke: https://www.fluke.com/en/product/industrial-imaging/sonic-industrial-imager-ii900

Swipeguide. (2013). Retrieved from https://www.swipeguide.com/

Ultrasonic leak detector. (2023). Retrieved from UE systems: https://www.uesystems.com/applications/ultrasonic-leak-detection/

US Department of Defence. (1980). *Procedures for performing a failure mode effects and criticality analysis.*

What is a cmms. (2013). Retrieved from IBM: https://www.ibm.com/topics/what-is-a-cmms

Wikipedia. (n.d.). *United Airlines Flight 232.*

www.ingramcontent.com/pod-product-compliance
Lightning Source LLC
Chambersburg PA
CBHW081058290526
45795CB00006B/1901